W9-BUW-998

Oxford University Press, Walton Street, Oxford OX2 6DP
Oxford is a trade mark of Oxford University Press

First published 1992
ISBN 0 19278131 6

Printed in Hong Kong

THE OXFORD BOOK OF Scarytales

Dennis Pepper

Oxford University Press
Oxford New York Toronto

CONTENTS

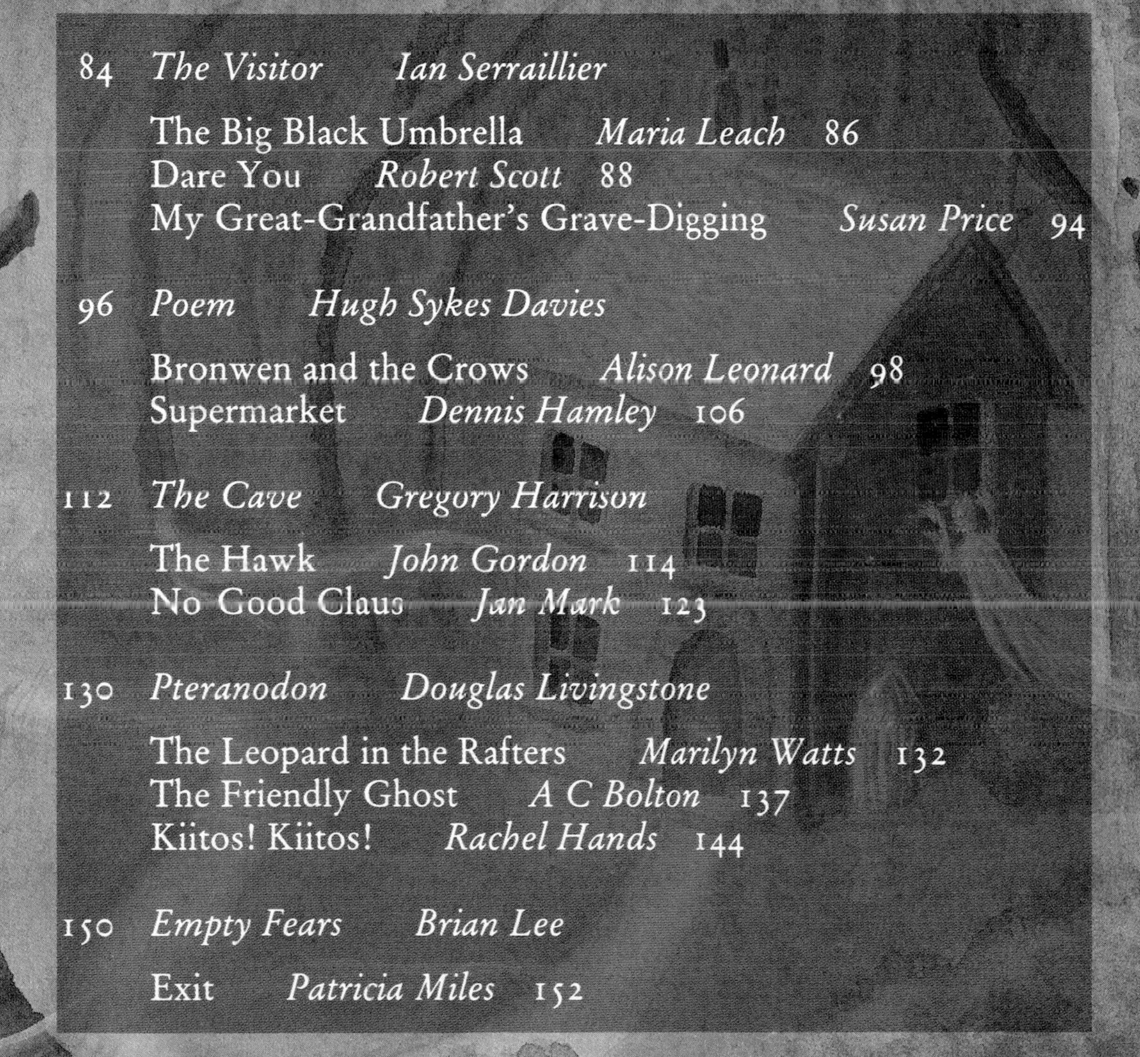

Scarytales . . .

You would expect a collection of scary tales to contain stories about the things people always think of as scary: ghosts and graveyards, witches, devils, monsters and 'things that go bump in the night'. They are here, of course. You will encounter the *real* ghost of Stonehouse Bay; discover what happened to Aunt Gertrude at Teppit's Pond; meet the Hairy Man; the Great Swallowing Monster and the Slitherydee; come upon creatures—tigers, lions, hawks, crows—that are not all they seem, and someone who isn't, surely, Santa Claus.

But not all these things will be scary for everyone who reads this book. After all, some people keep spiders as pets while others can't stand being in the same room as one. Moreover, in some of the stories the writer is, as it were, holding back so that you're amused almost by what happens even though the events weren't at all funny for the people in the story. We do see, though, how these people behave when faced with the things that frighten *them*.

Adults and children alike get a kind of satisfaction out of frightening themselves, and especially from reading frightening stories (which is rather safer than being involved with the real thing). We experience our fears at one remove; but an important question, for us as readers, as for the people in the stories, has to do with how we handle those fears. We may not, like the child in Dave Calder's poem, be able to ignore all the scary things and march off with the key to the castle. No doubt Justin was foolish to accept zit-faced Corkland's dare to adventure on to thin ice, but he meets his tiger—which has haunted his childhood—with courage and dignity. Gary, convalescent from a fever, goes to the supermarket with his mother and finds his nightmares coming true: 'he is *in* them and there is no escape.' But he struggles on, and is rightly proud of his achievement. Tiko and Marra, fleeing from the swallowing monster, are convinced of their cowardice and, at the end, confess their fears to the Chief. He understands, 'Well, *of course* you were scared! . . . But you came back, didn't you?'

Precisely. These children, and others too, are scared in ways that Dave Calder's child wouldn't even understand, but in the end they win through and go off with the key to their particular castle.

Dennis Pepper
29 January 1992

This is the Key to the Castle

Dave Calder

this is the key to the castle

this is the box
with rusty locks
that holds the key to the castle

this is the spider, huge and fat,
who wove her web and sat and sat
on top of the box
with rusty locks
that holds the key to the castle

this is the cellar, cold and bare,
dark as a grave, with nothing there
except the spider, huge and fat,
who wove her web and sat and sat
on top of the box
with rusty locks
that holds the key to the castle

this is the stair that rumbles and creaks
where every small step moans and squeaks,
that leads to the cellar, cold and bare,
dark as the grave, with nobody there
except the spider, huge and fat,
who wove her web and sat and sat
on top of the box
with rusty locks
that holds the key to the castle

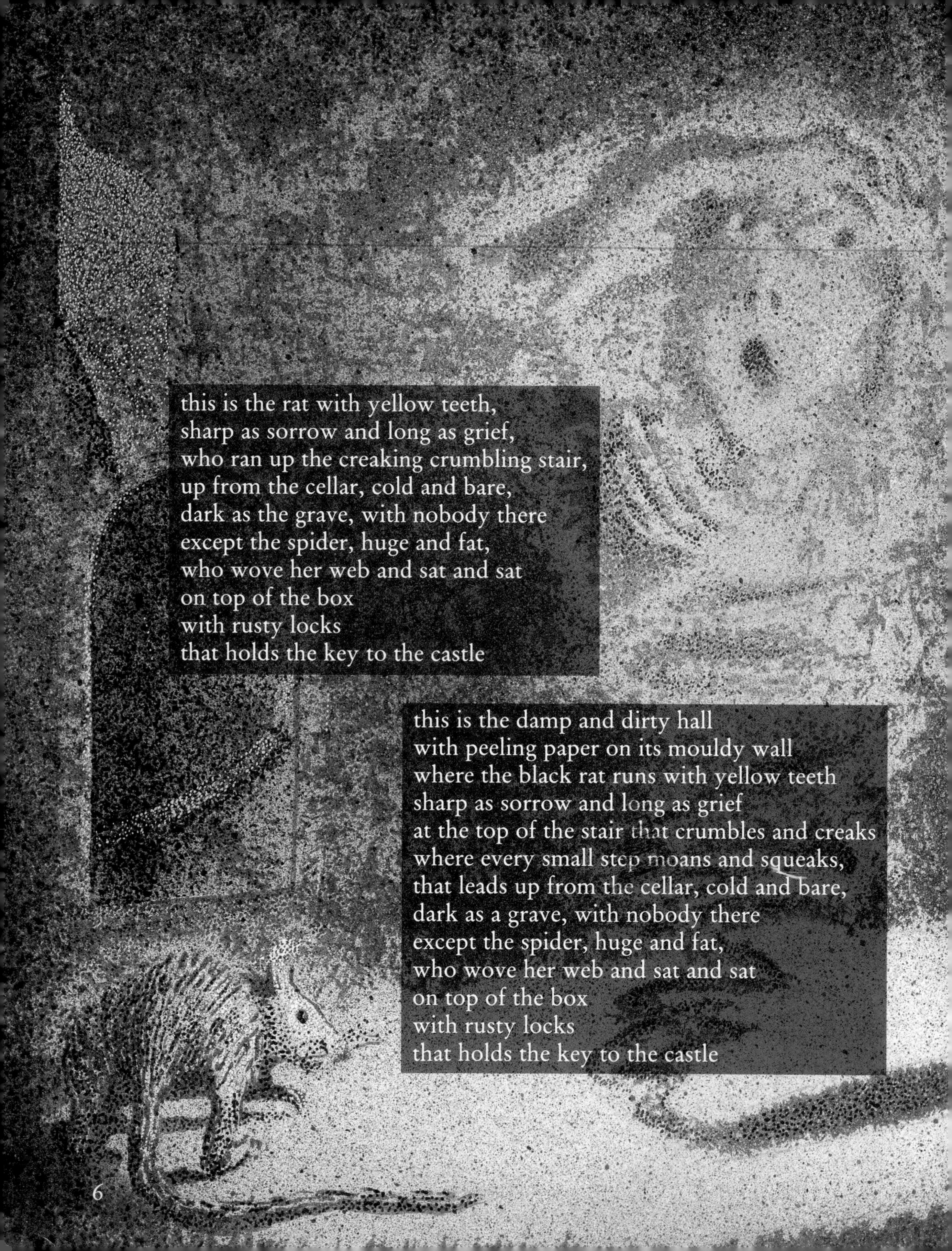

this is the rat with yellow teeth,
sharp as sorrow and long as grief,
who ran up the creaking crumbling stair,
up from the cellar, cold and bare,
dark as the grave, with nobody there
except the spider, huge and fat,
who wove her web and sat and sat
on top of the box
with rusty locks
that holds the key to the castle

this is the damp and dirty hall
with peeling paper on its mouldy wall
where the black rat runs with yellow teeth
sharp as sorrow and long as grief
at the top of the stair that crumbles and creaks
where every small step moans and squeaks,
that leads up from the cellar, cold and bare,
dark as a grave, with nobody there
except the spider, huge and fat,
who wove her web and sat and sat
on top of the box
with rusty locks
that holds the key to the castle

this is the ghost with rattling bones
carrying his head, whose horrible groans
fill the damp and dirty hall
with peeling paper on its mouldy wall
where the big black rat with yellow teeth
sharp as sorrow, long as grief,
runs to the stair that crumbles and creaks
where every small step moans and squeaks
that leads to the cellar, cold and bare
and dark as a grave with nobody there
except the spider, huge and fat,
who wove her web and sat and sat
on top of the box
with rusty locks
that holds the key to the castle

this is the child who came to play
on a rainy, windy, nasty day
and said BOO to the ghost who groaned in the hall
and SCAT to the rat by the mouldy wall
and went down the creaking crumbling stair
into the cellar, cold and bare,
and laughed at the spider, huge and fat,
and brushed off the web where it sat and sat,
and opened the box
with the rusty locks
and took the key to the castle

Dear Jane

Sheila Lavelle

Stonehouse Bay 10th June, 5pm.

Dear Jane,

What rotten luck you getting that sore throat at the last minute, when we'd been looking forward to school camp all this time! You must feel as sick as a parrot to be missing all the fun.

Why did you tell your mum? You might have known she'd make you stay in bed. Now I've got nobody to share my tent with, and I'll have to scoff that midnight feast all by myself.

Never mind. Like I promised, I'm going to write to you every single day.

Well, we finally arrived at about lunchtime, with Mr Harrison reading the map as usual. We only got lost four times, which isn't bad compared with last year when we went to Porthaddock and the teachers ended up not speaking to each other. This time they were as happy as hyenas, except for Mrs Grone. That woman's never happy about anything. If the dishiest bloke in the world gave her a kiss she'd ask him if he'd cleaned his teeth.

Anyway, we got the gear out of the bus and set up camp. It's a brilliant place for camping, with nothing except the sea and the birds and a crumbling mansion that's been empty for years and looks dead spooky with the windows falling out and brambles growing over it.

I dumped my tent in a good spot and went to collect the rest of my stuff. When I got back somebody had shoved my tent out of the way and was setting up hers in its place. No marks for guessing who. Yes, that awful Audrey Armpit.

So I pulled a few tent pegs out and her tent fell down on her head, and in no time we were rolling about pulling each other's hair out. Everybody cheered like mad until Mr Harrison dragged us apart. He made us put our tents up at opposite ends of the bay, and

Mrs Grone says if I cause any more trouble I'll get sent home. As if it was my fault! Never mind, at least I'm as far away as possible from Audrey Armpit, and that suits me fine.

I'll put this in the postbox now and write again tomorrow. We're having a barbecue tonight, with stories and a sing-song round the campfire like we did last year. I hope Mr Harrison hasn't brought his guitar.

I can smell the sausages already!

Love,

Sarah

Stonehouse Bay 11th June, 5pm.

Dear Jane,

What a shame you're not here! We've had a great day, swimming and sailing and fishing and grilling mackerel over the campfire, but it's not the same without you. I had to sail the dinghy with Freddie Milton and we capsized six times. He's got as much idea about sailing as a camel up an apple tree with its eyes shut.

The worst thing was having to sleep in my tent on my own. Especially after the creepy ghost stories round the campfire in the dark. Mr Harrison's was the creepiest, all about the ghost of Stonehouse Bay. It was enough to make your perm go straight, I can tell you. It was about this girl who lived in the big house who drowned herself because her boyfriend ran away with somebody else instead of coming in his boat to marry her like he'd promised. She walked straight out to sea in her nightie, and her dead body got washed up weeks later, all yellow and bloated and crawling with maggots. Now her ghost walks the beach on moonlit nights, weeping and moaning and trailing seaweed, and drowning itself all over again.

The boys said what a load of rubbish, but some of the girls believed every word. Mrs Grone said we shouldn't fill our heads with such nonsense, especially at bedtime, and Mr Harrison should know better.

Then when I crawled into my tent and put my feet in my sleeping-bag, something cold and wet and slimy started slithering around inside. I switched my torch on and you'll never guess what it was. Four dead mackerel, staring at me with reproachful glassy eyes as if it was my fault they'd been born fish instead of humming-birds. Of course I knew who'd put them there, and I'll get my own back as soon as I think of something horrible enough.

Must go now or I'll miss the post. I'll write again tomorrow. We're cooking beefburgers tonight, and I expect Mr Harrison will be telling more scary stories. I hope he does. At least it keeps him from playing the guitar.

Wish you were here.

Love,

Sarah

Stonehouse Bay 12th June, 5am.

Dear Jane,

It's just getting light and I'm writing this before the others get up because I'll have to post it without anybody seeing me. Mrs Grone says I've got to stay in my tent until they decide what to do with me, after the uproar I caused last night.

It started after we had the beefburgers and stuff for supper, and blooming awful it was too. I smothered everything in tomato ketchup but it was still like eating burnt socks. So while they were singing 'Michael Row the Boat Ashore' for the fifteenth time I went back to my tent for a bag of crisps.

While I was there I checked my sleeping-bag, and this time it was a dead hedgehog which she must have picked up off the road and it couldn't have been dead long because it was still hopping with fleas.

I chucked the poor squashed thing into the brambles and went back to the campfire, vowing I'd get my own back if it was the last thing I did. And you won't believe this but they were all begging Mr Harrison to tell the Stonehouse Bay ghost story all over again. And he did, even though Mrs Grone said we'd all have nightmares. It was when he got to the bit about how the ghost haunts the shore in her nightie with seaweed in her hair that I noticed Audrey Armpit get up and go sneaking off down the beach in the dark.

I knew she was up to no good, so I followed her. I didn't know what to think when I saw her collecting a great armful of seaweed and taking it back to her tent. Nobody needs that amount of seaweed to find out if it's going to rain, so I crawled up to her tent-flap and put my eye to the gap. There she was, sitting on her sleeping-bag, cutting a hole in her sheet with the scissors from her first-aid kit!

When she'd put her head through the sheet and draped it round her, she covered her face with ointment, also from the first-aid kit, and made it all yellow and shiny. I must be as thick as two short planks but it was only when she started fixing seaweed in her hair with hair-grips that it dawned on me what she was up to. I said a very rude word and crept away to my own tent, vowing to beat Audrey Armpit at her own game.

I was glad Mrs Grone had issued everybody with that first-aid kit. It didn't take long to make a nightie out of my sheet and smear my face with ointment, too. I knew Audrey couldn't do anything until everybody had gone to bed, so I waited until the camp was quiet and still. Then I slipped out of my tent and picked my way down the beach to the line of seaweed.

I just wish you could have seen me when I had draped my head with weed and tucked my torch into my belt so that the light shone into my greasy yellow face. I'd teach Audrey Armpit to play tricks on me, I told myself gleefully, and off I went towards her tent.

I only got halfway across the beach when I saw a figure in white coming towards me from the direction of the old house. Right, Audrey Armpit, I said to myself, now you're for it. And I stood there waving my ghostly arms about, trying not to giggle, and expecting her to scream the place down as soon as she saw me.

The figure came closer, making the most dreadful moaning sounds I've ever heard in my life. I saw its ghastly yellow face and its dripping wet nightie and the long strands of seaweed in its hair, and I thought, good grief, Audrey Armpit, you should go on the telly.

And now for the horriblest bit.

The figure walked straight past me and took no notice of me at all!

As it went past, the moon came out from behind a cloud and I saw the sunken dead eyes and the maggotty rotting cheeks. And then I went cold all over, for the figure carried on walking, straight down the beach into the sea. It waded in deeper and deeper, until with a last horrible cry that made my seaweed stand on end it disappeared under the waves.

The tide was coming in and the water was lapping round my feet but I was too terrified to move. All I could do was stand there with my eyes shut and my mouth open and scream my head off. And it never occurred to me that there was too much screaming going on for me to be doing it all myself. It was only when I heard people shouting that I opened my eyes and saw that only ten yards away yet another ghost in a long white nightie had its hands over its face and was screaming too.

The whole camp was soon running about in a panic. It was Mrs Grone who brought me to my senses, stomping over the wet sand in her slippers and her Thomas the Tank Engine pyjamas with her curlers in her hair.

'Sarah Jennings! Audrey Armstrong! Just what do you think you're playing at?' she bellowed, and it was a waste of time telling her what we'd both just seen because of course she didn't believe a word of it.

Can't write any more now. It's six o'clock and time to sneak out to the post.

I'll let you know what they decide to do with me.

Love,

Sarah

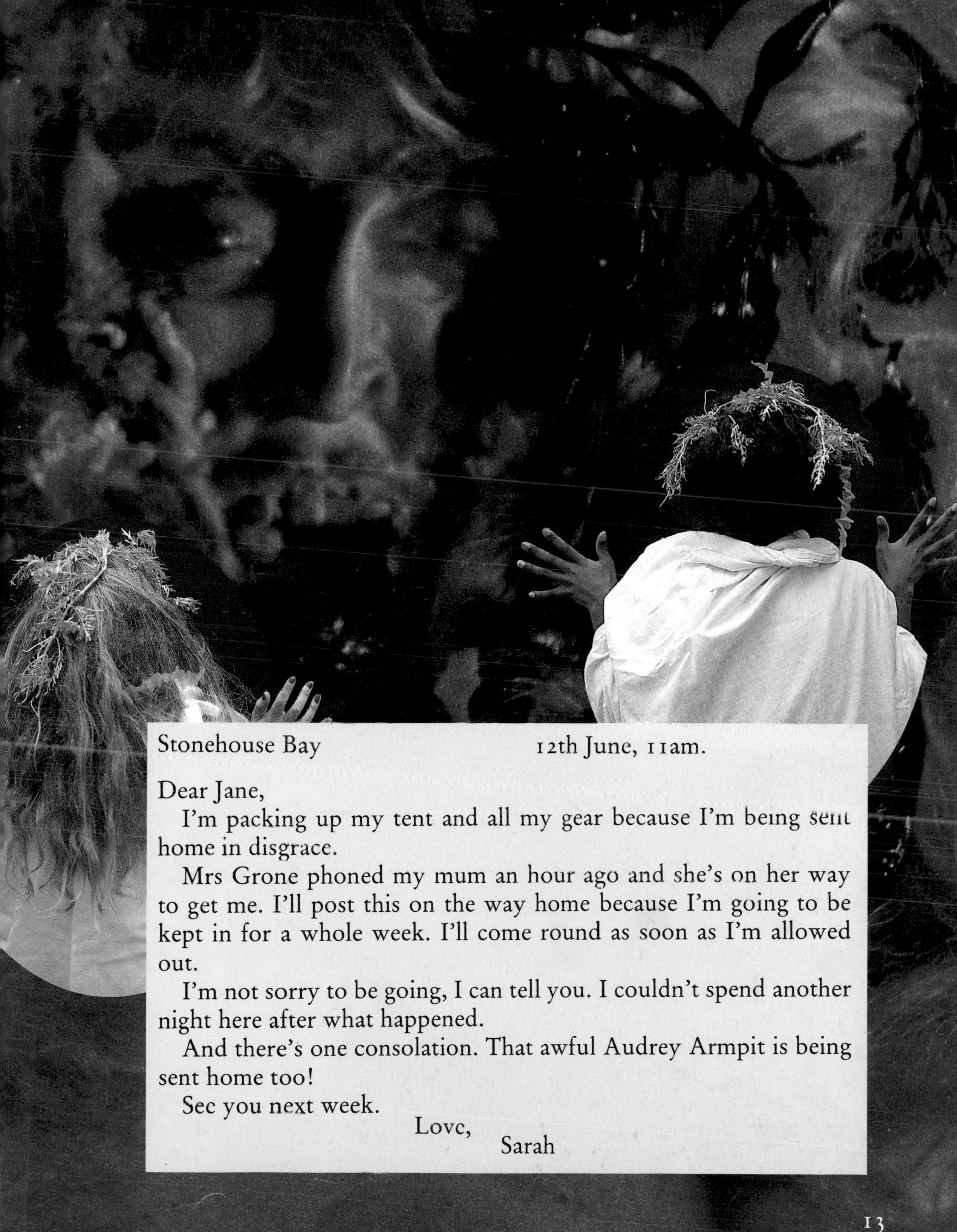

Stonehouse Bay 12th June, 11am.

Dear Jane,

I'm packing up my tent and all my gear because I'm being sent home in disgrace.

Mrs Grone phoned my mum an hour ago and she's on her way to get me. I'll post this on the way home because I'm going to be kept in for a whole week. I'll come round as soon as I'm allowed out.

I'm not sorry to be going, I can tell you. I couldn't spend another night here after what happened.

And there's one consolation. That awful Audrey Armpit is being sent home too!

See you next week.

Love,

Sarah

Slam and the ghosts

Kevin Crossley-Holland

'Night after night,' said Slam's mother. 'The boor! The great clod-hopper!'

'You can't talk to him without getting angry,' said Douglas. 'You can't even talk about him without getting angry. I'll talk to him.'

'I'll brain him! Bursting in at one o'clock night after night! Blundering about! Leaving his great hoof marks all over the house, the drunkard!'

'I'll talk to him,' said Douglas again.

'I don't know,' said his mother. 'You're so alike—always loyal to each other, always wanting to avoid a scrap. You and Slam, you look the same too.'

'I wonder why,' said Douglas.

'And so unalike . . . You work; you bring home the bacon. And Slam . . .'

'I know,' said Douglas quickly.

'It's the drink,' said Douglas's mother. 'It's wrecking him. Can't you get him off it?'

Secretly, Douglas agreed with his mother. *What Slam really needs*, he thought, *is a bit of a shock*.

Halfway between the pub and their cottage—and it was a couple of miles from one to the other, maybe a bit further—the lane passed under a very steep bank; and at the top of the bank was the old disused graveyard.

That same night, very late, Douglas pulled the white sheet off his bed. He let himself quietly out of the cottage and, under stars sharp as thorns, walked up to the graveyard. There Douglas wrapped the sheet around him and sat down on a gravestone right on top of the bank, overlooking the lane.

'This will cure him,' Douglas said to himself. 'Kill him or cure him. Poor old Slam!'

At much the same time as usual, Slam came staggering up the lane. His shoes were made of lead, and he was singing a wordless song.

When his brother was right beneath him, Douglas stood up and whoo-hooed at him.

'I know!' said Slam, and he added a great hiccup. 'You're the ghost! I know!'

Douglas whoo-hooed again and Slam peered up at the graveyard and tottered sideways.

'Two ghosts!' exclaimed Slam. 'There was only one ghost last night.'

Slowly Douglas turned round, and stared straight into two furious, glaring eyes.

Douglas started back and fell head first over the steep bank. He landed at his brother's feet and broke his neck. Poor old Douglas! That was the end of him.

Crossing over

Catherine Storr

If she hadn't been fond of dogs, she would never have volunteered for this particular job. When her class at school were asked if they would give up some of their spare time towards helping old people, most of the tasks on offer had sounded dreary. Visiting housebound old men and women, making them cups of tea and talking to them; she hadn't fancied that, and she wasn't any good at making conversation, let alone being able to shout loud enough for a deaf person to hear. Her voice was naturally quiet.

She didn't like the idea of doing anyone else's shopping, she wasn't good enough at checking that she'd got the right change. The check-out girls in the supermarket were too quick, ringing up the different items on the cash register. Nor did she want to push a wheelchair to the park. But walking old Mrs Matthews's dog, that had seemed like something she might even enjoy. She couldn't go every evening, but she would take him for a good long run on the Common on Saturdays, and on fine evenings, when the days were longer, she'd try to call for him after school some weekdays. She had started out full of enthusiasm.

What she hadn't reckoned with was the dog himself. Togo was huge, half Alsatian, half something else which had given him long woolly hair, permanently matted and dirty. Once, right at the beginning, she had offered to bathe and groom him, but Mrs Matthews had been outraged by the suggestion, was sure the poor creature would catch cold, and at the sight of the comb, Togo backed and growled and showed his teeth. It was as much as she could do to fasten and unfasten his leash, and he did not make that easy.

The early evening walks weren't quite so bad, because there wasn't time to take him to the Common, so he stayed on the leash all the time. Even then he was difficult to manage. He seemed to have had no training and he certainly had no manners. He never stopped when she told him to, never came when she called him, so that every Saturday, when she dutifully let him run free among the gorse bushes and little trees on the Common, she was afraid she might have to return to Mrs Matthews without the dog, confessing that he had run away.

Mrs Matthews did not admit that Togo was unruly and difficult to manage, any more than she would admit that he smelled. It was only a feeling that she shouldn't go back on her promise to perform this small service to the community that kept the girl still at the disagreeable task.

This particular evening was horrible. She'd been kept later at school than usual, and although it was already March, the sky was overcast, it was beginning to get dark, and a fine drizzling rain made the pavements slippery.

Togo was in a worse mood than usual. He had slouched along, stopping for whole minutes at lamp-posts and dustbins and misbehaving extravagantly in the most inconvenient places, in spite of her frantic tugs at the leash to try to get him off the pavement. He was too strong for her to control, and he knew it. She almost believed that he had a spite against her, and enjoyed showing that he didn't have to do anything she wanted, as if it wasn't bad enough having to go out in public with an animal so unkempt and anti-social.

They reached the zebra crossing on the hill. The traffic was moving fast, as it always did during the evening rush-hour. She would have to wait for a break before she could step off the pavement, especially as, in the half dark, she knew from her Dad's comments when he was driving, pedestrians on the road were not easy to see.

She stood still and dragged at Togo's lead. But Togo did not mean to be dictated to by a little schoolgirl, and after a moment's hesitation, he pulled too. He was off, into the middle of the on-coming traffic, wrenching at the leash, which she had twisted round her hand in order to get a better grip. She threw all her weight against his, but she was no match for him. She thought she felt the worn leather snap, she heard the sound of screaming brakes and some-one shouted. She had time to think, 'What am I going to say to Mrs Matthews?' before her head swam and she thought she was going to faint.

She found herself standing on the further side of the road. She saw a huddle of people, surrounding sta-tionary cars. Two drivers had left their vehicles and were abusing each other. As the crowd swayed, she saw the bonnet of a red car crumpled by its contact with the back of a large yellow van. She saw, too, a dark stain on the road surface. Blood. Blood made her feel sick, and her head swam again.

She hesitated, knowing that she ought to go among the watching people to make herself look, perhaps to try to explain how Togo had pulled, how she hadn't been strong enough to hold him back. Someone should be told whose dog he was. Someone would have to go and break the terrible news to Mrs Matthews.

As she was considering this, she heard the siren of a police car and the two-note call of an ambulance. She thought, 'Perhaps someone got badly hurt in one of the cars, and it's all my fault.'

Her courage evaporated, and she turned away from the accident and began to walk, on legs that trembled, up the hill towards her own home. She thought, 'I'll go and tell Mum.' But then she remembered how much Mrs Matthews loved horrible Togo, how she talked about him as her only friend, and how dreadful it was going to be for her to open her front door to find a policeman telling her that her dog was dead. Besides, the policeman might say that it was all her, the girl's, fault. She had to go first to Mrs Matthews's house, to break the news gently, and also to explain that she had tried her best to prevent the accident.

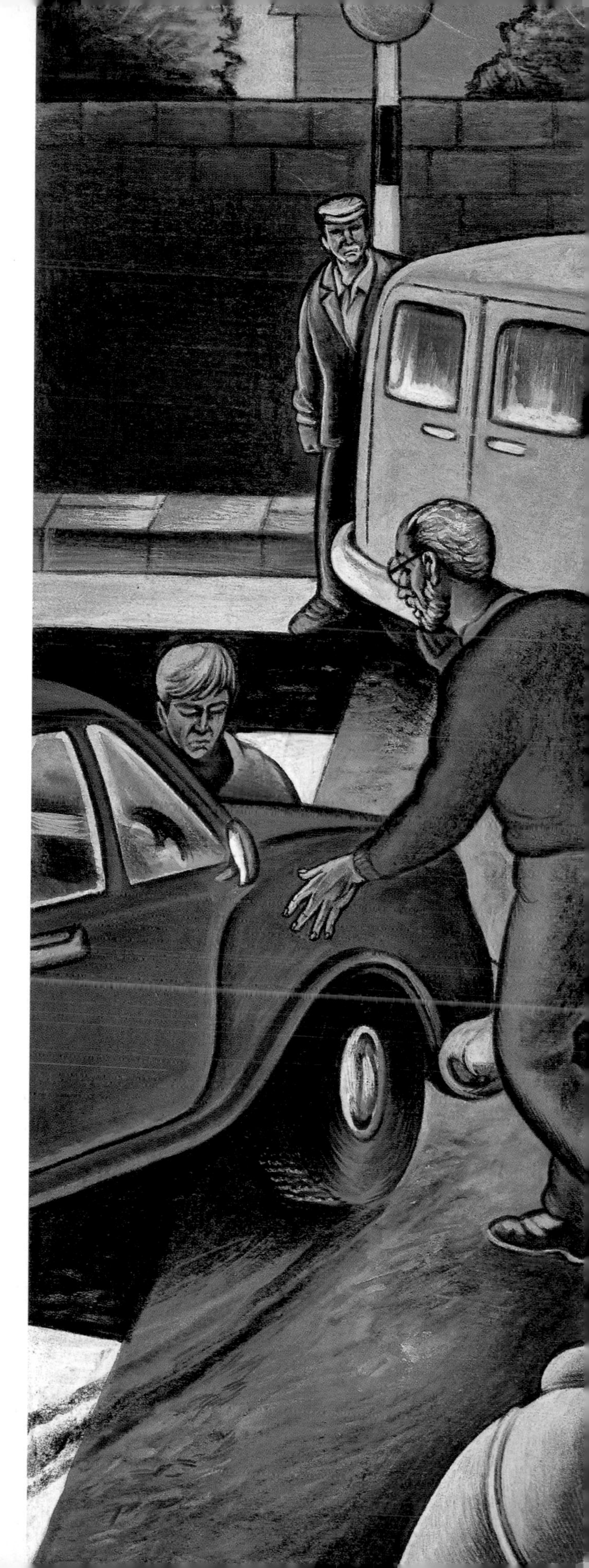

She found that she must have been walking really fast, which was surprising, considering how much she was dreading the ordeal in front of her. She had reached the grocer's and the newspaper shop at the top of the High Street almost before she'd realized.

She saw Sybil Grainger coming out of the newspaper shop, and she was ready to say, 'Hi!' and to pretend that there was nothing wrong, but luckily Sybil seemed not to have seen her. She turned the corner into Grange Road, relieved that she hadn't had to carry on a conversation. Grange Road also seemed shorter than usual; now she had to go along Fenton Crescent till she reached the small side street where Mrs Matthews lived, in one of the row of little old cottages known as Paradise Row.

Her heart beat furiously as she unlatched the small wooden gate and walked the short distance up to the front door, rehearsing exactly how to say what she had to. She lifted the knocker. As it came down on the wood, it made a hollow, echoing sound.

Extraordinary. From the other side of the door, she heard something very much like Togo's deep, menacing growl. She must be in such a state of nerves that she was imagining impossible things. Or perhaps when she felt faint out there in the road, she had fallen and hit her head and been concussed. She felt her scalp, under the straight, silky hair, but she couldn't find any tender spots. She waited. Mrs Matthews was arthritic and always took a long time to answer the door, and there was no hurry for the message she was going to receive.

Steps came slowly, dragging a little, along the passage. The door opened, and she braced herself for the shock she was about to administer and the scolding she was certainly going to receive.

But when Mrs Matthews looked out, she behaved in a very peculiar way. Instead of saying immediately, 'Where's Togo?' she asked nothing of her visitor, but bent forward and peered out, looking up and down the short row of cottages, as if she were searching for something or someone who might be coming or going in the street. Her head with its thinning grey hair was so close that the girl stepped back, opening her mouth to begin her explanation. But what she saw in the passage behind the old woman stopped her from uttering a sound.

At the further end of the passage was a dog. Togo. Togo, whole, apparently unharmed, his collar round his neck, and the end of the broken leash still attached, dragging behind him.

For a moment she thought he was going to spring forward and attack her. Then she saw that, instead, he was backing, shrinking as far away as he could get. He was making a curious noise, not a howl, nor a growl, but a sort of whine. She noticed that he was trembling. She had never seen Togo tremble before. He was showing whites round his yellow eyes and the short hair round his muzzle was bristling.

She started to speak. But Mrs Matthews appeared not to have heard her. She was turning back to calm the terrified dog. She was saying, 'Whatever's the matter with you, Togo? Think you're seeing a ghost?'

Miller's End

Charles Causley

When we moved to Miller's End,
Every afternoon at four
A thin shadow of a shade
Quavered through the garden-door.

Dressed in black from top to toe
And a veil about her head
To us all it seemed as though
She came walking from the dead.

With a basket on her arm
Through the hedge-gap she would pass,
Never a mark that we could spy
On the flagstones or the grass.

When we told the garden-boy
How we saw the phantom glide,
With a grin his face was bright
As the pool he stood beside.

'That's no ghost-walk,' Billy said,
'Nor a ghost you fear to stop—
Only old Miss Wickerby
On a short cut to the shop.'

So next day we lay in wait
Passed a civil time of day,
Said how pleased we were she came
Daily down our garden-way.

Suddenly her cheek it paled,
Turned, as quick, from ice to flame.
'Tell me,' said Miss Wickerby,
'Who spoke of me, and my name?'

'Bill the garden-boy.'
She sighed,
Said, 'Of course, you could not know
How he drowned—that very pool—
A frozen winter—long ago.'

A Change of Aunts

Vivien Alcock

Everyone knows the pond in Teppit's Wood is haunted. A young nursemaid once drowned herself there. She had done it early one evening, with the sun sinking in the red sky, and the smoke from the burning house drifting through the trees.

They say she had slipped out to meet her sweetheart, and left the two little children alone, with the fire blazing behind its guard in the nursery grate. Burnt to cinders they were, the poor little ones, and the young nursemaid, mad with the guilt and grief of it, had done away with herself.

But she still can't rest, the tale goes; and at sunset, you'll see the smoke drifting through the trees, though a hundred years have passed since the big house burned down. Then, if you're wise, you'll run! For that's when the poor crazed ghost rises up, all wet from the dark pond, and goes seeking the dead children. Searching and searching all through the woods for the little children . . . *Take care she doesn't get you!*

Meg Thompson, who was eleven, thought perhaps she was too old to believe in ghosts. Her brother William believed in them, but he was only eight. Aunt Janet seemed to, but perhaps she was only pretending, just to keep William company, so that he need not feel ashamed.

Even in full daylight, Aunt Janet would hold their hands and run them past the pond, chanting the magic charm:

> 'Lady of the little lake,
> Come not nigh, for pity's sake!
> Remember, when the sun is high,
> We may safely pass you by.'

And they would race up the hill through the trees, until they arrived home, laughing, breathless and safe.

They loved Aunt Janet, who had looked after them ever since their mother had died. Unfortunately, a neighbour's brother, come visiting from Australia, loved her too, and carried her back to Adelaide as his bride.

That was when Aunt Gertrude came. She was as different from Aunt Janet as a hawk from a dove. Thin and hard and sharp, she seemed to wear her bones outside her skin and her eyes on stalks. She could see dirty fingernails through pockets, smuggled bedtime cats through blankets, and broken mugs through two layers of newspaper and a dustbin lid.

'I'm up to all your tricks,' she told them, with a smile like stretched elastic.

She only smiled when their father was in the room. There were many things she only did when he was there, such as calling them her dears, and giving them biscuits for their tea, and letting them watch television. Just as there were many things she only did when their father was out, such as feeding them on stale bread and marge, slapping and punching them, and locking them in the cellar as a punishment.

They did not mind being shut in the cellar. They played soldiers with the bottles of wine, or cricket with a lump of coal and a piece of wood. Or they sat on empty crates and planned vengeance on Aunt Gertrude.

'I'll get a gun and shoot her,' William said. 'I'll cut her up into little pieces with the carving knife and feed her to Tiddles.'

'You'd only get sent to prison,' Meg objected. 'I'm going to write a letter to the Child Welfare and tell them about her, and they'll put *her* in prison.'

'They won't believe you,' William said, 'any more than Dad does.'

Meg was silent.

'Why doesn't Dad believe us?' William asked.

'Because she's always nicer to us when he's here. Because she doesn't hit us hard enough to leave bruises. Because she's told him we're liars.' Meg hesitated, and then added slowly, 'And because he doesn't *want* to believe us.'

'Why not?'

'She's our last aunt. If she went, he wouldn't know what to do with us. He might have to send us away, and that would be worse.'

William looked doubtful, but before he could say anything, there was the sound of a door shutting upstairs.

'She's back! Look sad, William,' Meg whispered. They did not want Aunt Gertrude to find out they did not mind being locked in the cellar. She'd only think of another punishment. One that hurt.

'Meg,' William whispered anxiously, 'you haven't told her about the haunted pond, have you?'

Meg shook her head.

'She'd take me down there, I know she would. At sunset,' William whispered, his eyes huge with fear. 'At sunset, when it's dangerous to go.'

'I won't let her,' Meg said.

In September, their father had to go to Germany for a month on business. They both cried when he left, and this made Aunt Gertrude angry. As a punishment, she sent them to bed without supper, locking their rooms so that they could not sneak down in the night to steal food from the kitchen.

'I'm up to all your little tricks,' she told them.

They were so hungry the next day that they were almost glad it was Wednesday. For every Wednesday, Aunt Gertrude took them to tea with a friend of hers, who lived in Eggleston Street, three miles away by road and no buses. Mrs Brown was as square as Aunt Gertrude was angular, but otherwise seemed to be made of the same material. Granite. But at least they got sandwiches and cake there, and could shut their ears to the insults the two women aimed at them.

'The trouble I've had with them,' Aunt Gertrude started off.

'I don't know what children are coming to, I'm sure,' Mrs Brown agreed. And they went on and on until at last it was time to go.

The walk back was all up hill. Usually Aunt Gertrude would stride ahead, and shout at the children when they lagged behind. They never complained when their legs ached and blisters burst on their heels. They did not want Aunt Gertrude to find out about the short cut through Teppit's Wood. But this Wednesday, as they were getting ready to go, Aunt Gertrude said that she was tired.

'Looking after these two wears me out. I must tell John he'll have to buy me a car. It's a long walk back up Eggleston Hill. . .'

'Up Eggleston Hill?' Mrs Brown repeated, surprised. 'Don't you take the short cut through the wood?'

The children looked at each other in alarm.

'What short cut?' Aunt Gertrude demanded. 'I didn't know there was a short cut. Nobody told me. . .' Her eyes looked round for someone to blame, and found the children: 'Did you know about the short cut?' she asked angrily.

'Of course they knew. Everyone knows,' Mrs Brown said. She looked at Meg and William and smiled nastily. 'Don't tell me you're afraid to pass the haunted pond? I thought only babies were afraid of ghosts!' The sinking sun, shining through the window, flushed her face as if with wine. 'Never mind,' she said, her voice as falsely sweet as honey from a wasp, 'I'm sure your dear Aunt Gertrude will cure you of such silly fancies.'

William clutched hold of Meg's hand.

'I'm not going through the wood! I'm not! You can't make us! Not at sunset!'

Meg put her arms round him. She could hear Mrs Brown telling Aunt Gertrude about the ghost of the young nursemaid, and Aunt Gertrude laughing scornfully.

'So you're frightened of ghosts, are you?' she said to the children, after they had left the house. 'You'd let your poor aunt walk two unnecessary miles because of some stupid old wives' tale. Your poor aunt who works so hard while you spend all day playing! I'll soon see about that.'

She grabbed them each by a wrist with her hard fingers, and dragged them down the path into the woods. The trees closed round them in a dark, whispering crowd, seeming to murmur, 'The sun is setting . . . keep away, keep away!'

William began to struggle and kick. Aunt Gertrude let go of Meg and hit William so hard that he was knocked right off the path. He fell into a deep drift of dead leaves, which rose up like brown butterflies and settled down on him, as he lay whimpering.

Meg ran to comfort him. 'You'll have a bruise,' she whispered softly. 'You'll have a big bruise to show Dad when he comes back.'

He smiled through his tears.

'What's that? What are you two plotting?' Aunt Gertrude asked sharply. 'Any more nonsense out of you, and there's plenty more where that came from. Well? Are you going to behave?'

She stood over them, tall and thin and hard as an iron lamp-post, with the setting sun seeming to glow redly in her hateful eyes.

'Meg,' William whispered, his arms round her neck, 'I think she's a witch. Don't you? Meg, d'you think she's a witch?'

'No,' Meg whispered back, more decidedly than she felt. 'Come on, we'd better do what she says. Don't be frightened. I'll look after you, William.'

So they walked down into the sighing woods. Their aunt marched behind them, throwing a long shadow that struck at their feet. William held tight to Meg's hand, and as soon as the dark pond came into sight, they began to chant under their breaths the words of the magic charm.

> 'Lady of the little lake,
> Come not nigh, for pity's sake!
> Remember, when the sun is high . . .'

'What are you two whispering about?' Aunt Gertrude demanded.

'Nothing,' they answered.

For it was no good, the magic charm. It only worked in daylight, when the sun was up. Now the sun had fallen into the trees, and the sky was on fire.

'Look!' William whispered.

Between the trees, pale wisps of smoke came curling and creeping over the ground, like blind fingers searching . . .

'It's the smoke! Meg, it's the smoke!' William screamed.

Aunt Gertrude grabbed his shoulder and shook him.

'Stop that din! Making an exhibition of yourself! It's only mist rising from the water. Come, I'll show you.' She started dragging William towards the pond. Meg grabbed him by the other arm, and for a moment they pulled him between them, like a cracker. Then Aunt Gertrude hit Meg hard on the ear, and Meg let go, putting her hands to her ringing head.

Aunt Gertrude forced William to the very edge of the dark pond.

'There! Look down, there's nothing there, is there, you stupid little coward? Answer me! There's nothing there, is there?'

She was looking at William as she spoke. She did not see what both the children saw. She did not see what rose out of the pond behind her.

It was something dark and wet, a figure of water and weeds.

Green mud clung like flesh to its washed bones. A frog crouched like a pumping heart in its cage of ivory. Its crazed eyes, silver as the scales of fishes, glared down at Aunt Gertrude as she hit the terrified boy. It reached out . . .

Aunt Gertrude screamed.

William pulled away from her and ran. Blind with fear, he raced past Meg without seeing her, and disappeared into the trees.

Meg could not move. She crouched down on the damp, leafy ground and watched in terror. Dark water was torn from the pond in creamy tatters as the two figures struggled together, the screaming aunt and the other one, all water and weed and bone. Its silver eyes glinting, it fastened its ivory fingers like combs into Aunt Gertrude's hair. Down, down they sank in a boil of bubbles.

'Meg! Meg!' William's voice called from among the trees, and Meg, as if released, leaped to her feet and ran after him, leaving Aunt Gertrude in the pond.

William had fallen over. His knee was bleeding, his bruised face wet.

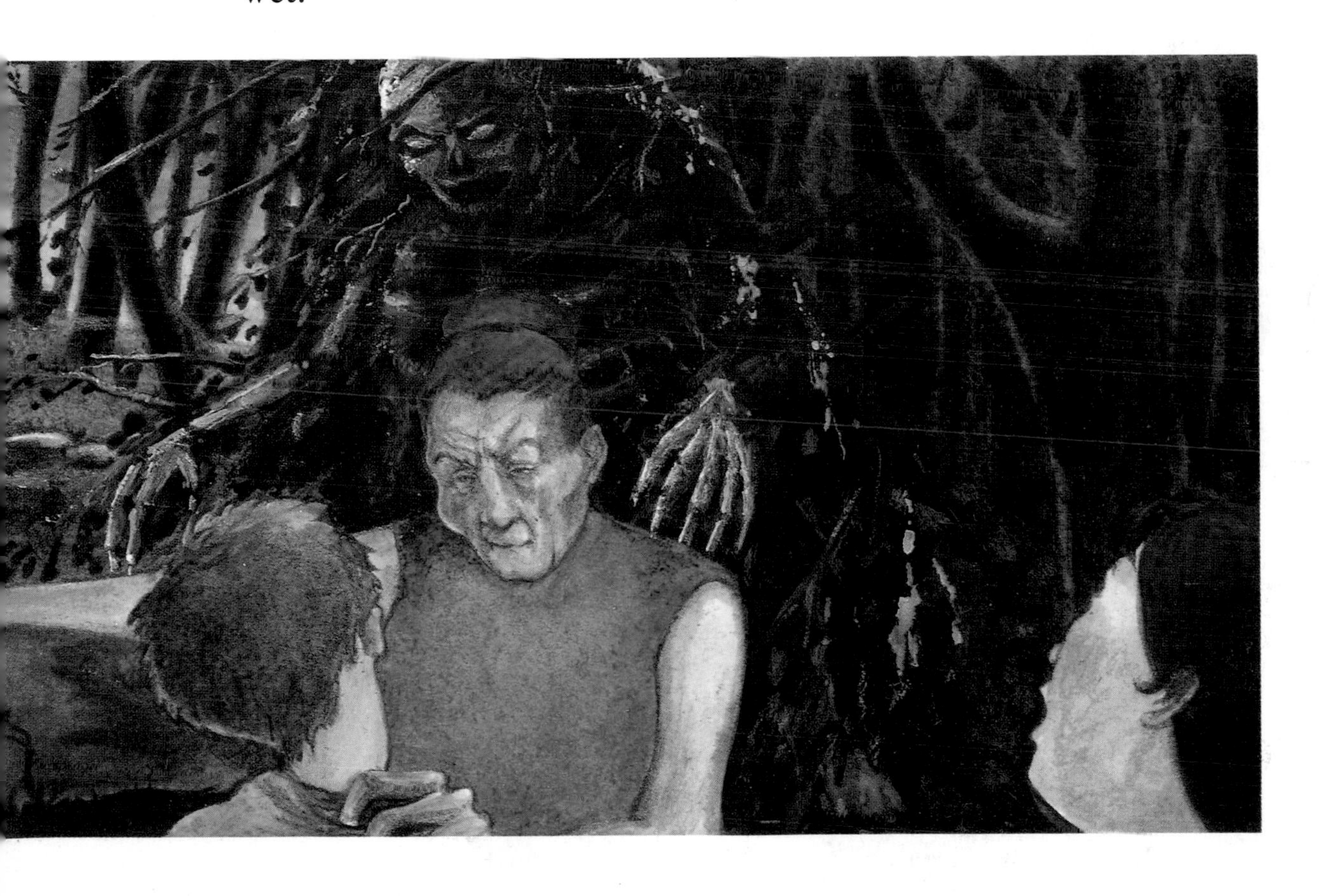

'Come on, come on, hurry!' Meg said, catching hold of his hand and dragging him after her.

For there was someone following. Running through the trees behind them, twigs snapping, leaves crunching under invisible feet.

'Run, William, faster, faster!' Meg cried.

'I can't!'

'You must! Run, William, run!'

It was nearer now, and nearer, following fast, bounding in huge leaps over the rotting branches and white nests of toadstools.

'Faster!' Meg cried, looking fearfully over her shoulder at the shaking bushes, not seeing the twisted root that caught at her feet. She fell, bringing William down with her.

Aunt Gertrude burst through the bushes.

How strange she looked! She had run so fast that the clothes had dried on her body, and her cheeks were pink. Her hair, loosened from its tight knot, was tumbled and tangled about her head.

The children cowered away from her as she came up and knelt down beside them.

'Are you all right, my little dears?' she asked softly. (*Dears*?) 'That was a nasty tumble! Why, you're shivering, Miss Margaret! And Master William, you've cut your poor knee.' (*Miss*? *Master*?) 'If you're a brave boy and don't cry, I'll give you a piggy-back home, and there'll be hot chocolate and cherry cake by the nursery fire.'

They stared at her, trembling. The look in Aunt Gertrude's eyes was soft and kind. The smile on Aunt Gertrude's mouth was wide and sweet. What was she up to? What cruel trick was she playing now?

They were silent as Aunt Gertrude carried William up the hill to their home. There, as good as her word, she gave them hot chocolate and cake, and sat them on the sofa while she bathed William's knee.

When she had finished, she stood up and gazed at the empty grate in the living room, while they watched her silently. Then she left the room. They sipped their hot chocolate, sitting side by side, listening. They could hear her going from room to room all over the house, as if looking for something.

'What's she up to?' William whispered.

'I don't know.'

'Did you see it? Did you see it . . . in the pond?'

'Yes.'

'What happened, Meg?'

'Aunt Gertrude fell in,' Meg said, and shivered.

'Why is she so . . . so different?'

'I don't know.'

'I wish Daddy were back,' William said, and his lip quivered. Meg put her arm round him, and they were silent again, listening to the footsteps going round and round the house, slowly, uncertainly, as if Aunt Gertrude had lost her way.

There was no doubt that Aunt Gertrude was a changed woman since she had fallen into the pond. Perhaps the water had washed the nastiness out of her. The house had never been so bright and cheerful. Their meals had never been so delicious. She made them apple pie and cherry cake, and let them lick out the bowls. She played leap frog with them in the garden, and never minded running after the balls at cricket. She told them bedtime stories and kissed them good-night.

William started calling her Aunt Trudie, and would often hold her hand, taking her to see some treasure; a large snail with a whirligig shell, a stone with a hole right through the middle or a jay's feather. Meg followed them silently, watching and listening. Once, when William did not know she was behind them, she heard him say:

'Aunt Trudie, you mustn't call us Miss Margaret and Master William, you know.'

'Should I not, Master William?'

'No. Just plain Meg and William.'

'William, then.'

'That's better. And when Daddy comes home on Saturday, you must call him John. Can you remember that?'

She smiled and nodded.

'Don't worry,' he said. 'I'll look after you, Aunt Trudie.' Then he caught sight of Meg behind them, and said quickly, 'We're just playing a game. Go away, Meg! We don't want you!'

'Now, Mas . . . Now, William, that's no way to speak to your sister,' Aunt Trudie said gently. 'Of course we want her.' She smiled at Meg. 'We are going to see the kittens next door. Come with us, Meg.'

Meg shook her head and walked back to the house. She went up to Aunt Gertrude's bedroom and looked round. It was bright and clean, and there were flowers on the dressing-table. There was no smell, no sense of Aunt Gertrude in it anywhere. It seemed like another person's room. Meg sat down on the bed, and thought for a long time.

Aunt Trudie found her there, when she came in from the garden, flushed and laughing. She hesitated when she caught sight of Meg, then called over her shoulder, 'Just a moment, William! Wait for me in the garden.'

Then she shut the door and leaned against it, looking gravely and kindly at Meg.

'Will you be staying with us long?' Meg asked politely.

'As long as ever you want me to,' was the answer.

There was a short silence. Then Meg jumped to her feet and put her arms round the woman.

'We don't want you to go, Aunt Trudie,' she said. 'We want you to stay with us for ever.'

It was three years before Meg ventured once more into Teppit's Wood. She went in broad daylight, when the sun was high. It was curiosity that took her there, down the winding path to the dark pond at the bottom. It was a warm day and birds were singing in the trees. The pond looked peaceful. There was frog-spawn in the brown water, leaves floated on the surface like little islands, and a water-boatman sculled across, leaving a silver wake behind him.

Meg stood a safe distance away and waited.

Bubbles began to disturb the quiet water. Small fish darted away and hid under the weeds. Now a scum of mud and filth rose slowly up from the bottom of the pond. It spread round a clump of frog-spawn, which shook and seemed to separate, and then reform into the shape of a hideous, scowling face.

As she watched, Meg thought she heard, faintly, a familiar voice.

'Meg! Get me out! Get me out this minute! She's stolen my body, that wretched servant-girl! Meg, if you bring her down here, I'll give you a penny. I'll give you chocolate biscuits every day. And roast beef! Just bring her down here and push her in! Meg, I'll never hit you again, I promise, I promise, promise. . .'

'Goodbye, Aunt Gertrude,' Meg said firmly, and left. That was the last time she ever walked in the woods round Teppit's pond.

Tiger in the Snow

Daniel Wynn Barber

Justin sensed the tiger as soon as he reached the street. He didn't see it, or hear it. He simply . . . sensed it.

Leaving the warm safety of the Baxters' porch light behind him, he started down the sidewalk that fronted State Street, feeling the night swallow him in a single hungry gulp. He stopped when he reached the edge of the Baxters' property line and looked back wistfully toward their front door.

Too bad the evening had to end. It had been just about the finest evening he could remember. Not that Steve and he hadn't had some fine old times together, the way best friends will; but this particular evening had been, well, magical. They had played *The Shot Brothers* down in Steve's basement while Mr and Mrs Baxter watched TV upstairs. When the game had been going well and everything was clicking, Justin could almost believe that Steve and he really were brothers. And that feeling had never been stronger than it had been this evening.

When Mrs Baxter had finally called down that it was time to go, it had struck Justin as vaguely strange that she would be packing him off on a night like this, seeing how he and Steve slept over at one another's homes just about every weekend. But this evening was different. Despite the snow, home called to him in sweet siren whispers.

Mrs Baxter had bundled him up in his parka, boots, and mittens, and then, much to his surprise, she had kissed his cheek. Steve had seen him to the door, said a quick goodbye, then hurried away to the den. Funny thing, Steve's eyes had seemed moist.

Then Justin had stepped out into the night, and Mrs Baxter had closed the door behind him, leaving him alone with the dark and the cold and . . . the tiger.

At the edge of the Baxters' property, Justin glanced around for a glimpse of the beast; but the street appeared deserted save for the houses and parked cars under a downy blanket of fresh snow. It was drifting down lazily now, indifferent after the heavy fall of that afternoon. Justin could see the skittering flakes trapped within the cones of light cast by the street lamps, but otherwise the black air seemed coldly empty. The line of lamps at every corner of State Street gave the appearance of a tunnel of light that tapered down to nothingness; and beyond that tunnel, the dark pressed eagerly in.

For a moment, Justin felt the urge to scurry back to the Baxters' door and beg for sanctuary, but he knew he should be getting home. Besides, he wasn't some chicken who ran from the dark. He was one of the Shot Brothers. Rough and ready. Fearless. Hadn't he proven that to stupid Dale Corkland just the other day? 'You scared?' old zit-faced Corkland had asked him. And Justin had shown him.

At the corner, Justin looked both ways, although he knew there wouldn't be many cars out on a night like this. Then he scanned the hedges along a nearby house, where dappled shadows hung frozen in the branches. Excellent camouflage for a tiger—particularly one of those white, Siberian tigers he'd read about.

He kept a close eye on those hedges as he crossed the street. Snow swelled up around his boots and sucked at his feet, making it impossible to run should a tiger spring from behind the mailbox on the far corner. He stopped before he reached that mailbox, listening for the low blowing sound that tigers sometimes make as they lie in ambush. But all he heard was the rasping of his own breath ('You scared?') Yes. Tigers were nothing to be trifled with. They were as dangerous as the ice on Shepherd's Pond.

Justin had stared at that ice, thinking about the warm weather they'd had the past week. Then he had looked up at Dale Corkland's face, three years older than his and sporting a gala display of acne. 'You scared?' And Justin had shown him.

But that was then and this was now; and weren't tigers more merciless than ice? Oh, yes indeed.

Justin gave himself a good mental shaking. He tried to summon those things his father had told him at other times when this tiger-fear had come upon him. (*Don't be such a baby.*) At night, when he would awaken screaming after a tiger nightmare. (*It was only a dream.*) Or when he felt certain that a tiger was lurking about the basement. (*There are no tigers in the city. You only find tigers in the zoo.*)

Wrapping himself snug in these assurances, Justin tramped past the brick retaining wall at the corner of State and Sixteenth without so much as a glance toward the spidery line of poplars where a tiger might be hiding. He rounded the corner and marched on. Heck, he had walked this way dozens of times. Hundreds, maybe.

But tonight the usually comfortable features seemed alien and warped out of reality under the snow, and finding himself in this strange white landscape, Justin suddenly felt the tiger-fear return. It bobbed up and down within him until he could almost feel the tiger's nearness, so close that the hot jungle breath seemed to huff against his cheek.

He was half-way down the block when he saw a shadow slip effortlessly from behind the house two doors up. It seemed to glide dream-like across the snow, then disappear behind a car parked in the driveway. It was just a shadow, but before it had vanished, Justin thought he caught a hint of striping.

There are no tigers in the city.

Justin watched and waited—waited for whatever it was to show itself. He even considered turning back, rerouting around Rush

Street, but that would put *it* behind him.

Come on, he scolded himself. You only find tigers in India. Or the zoo. *Or behind parked cars.* Nonsense. Tigers don't stalk kids from behind parked cars in the middle of an American city. Only little kids let themselves be scared by shadows in the night. Not one of the Shot Brothers. Not a kid who had dared the ice on Shepherd's Pond. Not a kid who was only two years away from attending Rathburn Junior High, where you get to keep your stuff in your own locker and change classrooms every hour and eat your lunch out on the benches. Kids at Rathburn didn't go whimpering and whining because they saw a shadow in the snow—probably thrown by a branch moving in the wind.

But there is no wind tonight.

Justin swallowed hard, then started forward. He walked slowly, never shifting his gaze from the tail light of that parked car. If only he could see around it without getting any nearer. If something were crouching back there, it would be on him before he could cover the first five feet. And then . . .

. . . teeth and claws, tearing and slashing.

You scared?

You bet.

When he had drawn even with the driveway across the street, Justin stopped. Two more steps, maybe three, and he would see if his father and the kids at Rathburn Junior were right, or if tigers do indeed lie in wait on winter streets. Of course, there was still time to turn back.

Perhaps it was the idea of turning back that propelled him forward. If he were to retrace his steps, he would never know; but if he looked and saw no tiger behind that car, then the tiger-fear would be banished, and he wouldn't see them anywhere. Not in bushes. Not behind trees. Not between houses. Just three steps, and he could lay tigers to rest forever.

Justin took those three steps the way he had walked out on to the ice on Shepherd's pond. Old zit-faced Corkland had dared him, and he had faced it.

One—two—three.

He turned and looked.

Nothing. Nothing behind that car but an old sledge lying on its side. No tigers. No lions, bears, werewolves, or boogie-men. Just an old sledge. His father had been right all along.

He covered the last block and a half with steps as light and carefree as those of a June day, when the air smelled of new-mown grass and the sun baked your skin brown. But, of course, it wasn't June, and as he sprinted up his porch steps Justin realized that he had reached home without a moment to spare. He could scarcely see his breath at all. Much longer out in the icy cold and he thought his lungs might have frozen solid.

As he stepped into the familiar warmth of his own house, he heard voices coming from the living room. It sounded as though his folks were having a party, although the voices seemed rather subdued—much the way they sounded on bridge nights when the evenings began quietly, but noisied up as the hours grew old.

Justin tip-toed down the hall, thinking it wise not to interrupt. And as he passed the living room, he caught a snatch of conversation. It was a man speaking, '. . . bound to happen eventually. They should have put up a fence years ago. I've a good mind to . . .'

'Oh, for God's sake, Gordon,' a woman said. (It sounded like Aunt Phyllis.) 'This isn't the time.'

That was all he heard before hurrying to his room.

When he flipped on the light, he was greeted by all the treasures which reflected his short life in intimate detail. The Darth Vader poster, the Packers pennant, the Spitfire on his dresser, the bedspread decorated in railroad logos.

And one new addition, sitting in the corner on great feline haunches.

For the briefest instant, Justin felt the urge to run—to flee into the living room and hurl himself into his mother's arms, as he had done so many times in the past. But as he stared transfixed into the tiger's huge, emerald eyes, he felt the fear slipping from him like some dark mantle, to be replaced by the soft and gentle cloak of understanding.

'It's time to go, isn't it?' he said in a voice that was low, but unwavering.

The tiger's eyes remained impassive, as deep and silent as green forest pools. Warm pools that never froze over, the way Shepherd's Pond did.

In his mind, Justin heard again the pistol crack of ice giving way beneath him, and he felt the chill water closing over his head. It really hadn't hurt that much, not the way he would have thought. Not much pain, just a moment of remorse when he realized he wouldn't be seeing his folks anymore—or Steve . . .

. . . had it all been a dream, this last wonderful evening together with Steve? Would Steve even remember?

Justin looked at the tiger, searching its peaceful face for the answer; but those fathomless eyes kept their secrets.

'Did you follow me tonight?' Justin asked.

Whiskers twitched as the tiger's muzzle wrinkled into a slight grin.

'Yes,' Justin said softly. 'I thought it was you. You've been following me all my life, haven't you?' He turned to close his bedroom door, and when he turned back, the tiger was crouching to spring.

Camilla

Adèle Geras

An interview with Monica Bridges

(author of *Ghoulies and Ghosties* and *Shades of Darkness*)
by Lynn Michael

L.M: I hope you don't mind if I use this tape recorder. I'm doing this interview for the Author Corner of our school magazine. In the back of your books, it says you spent some time abroad as a child. Where did you live?

M.B: I lived in a town called Jesselton, in North Borneo. Borneo is an island, across on the other side of the world, quite near China. It was a simply beautiful place. I don't think I realized how lovely it was when I was a girl. You take the place where you live for granted. I think I thought that everyone had a turquoise sea at the bottom of their garden. We also had a wonderful mountain called Kinabalu behind our house. It was purple sometimes, and sometimes green and at night it was black, outlined against the blue of the sky.

L.M: What was your school like?

M.B: It was a hut made out of palm leaves and bamboo standing in a field which became a swamp whenever it rained.

L.M: Did you like going to school?

M.B: Yes, because that was where all the children went. That was where you met your friends. There were only about thirty children. If you were over eleven, you were sent to boarding-school in England. Some of us were 'big ones' aged eight or nine or ten and the rest were 'little ones' who concentrated on learning how to read and write. We had one teacher in charge of all of us. She was called Mrs Ashton. She wasn't a qualified teacher, I don't think, but we all liked her, us big girls, because she wore pretty dresses with very full skirts, and because she used to tell us all about where and when she had met her husband. He was called Brian, and he worked in the police force. When he brought her to school every morning in the car, he kissed her before she got out. He always wore a nicely-pressed khaki uniform. My friend Sarah and I hid behind a tree and watched them kissing whenever we could.

L.M: Was Sarah your best friend?

M.B: I suppose so. I think so. Certainly she was my everyday, bread-and-butter sort of friend. Our parents did a lot of things together, so we saw a lot of one another. I liked her. I really did like her. I suppose she was my best friend, certainly until Camilla came.

[*At this point there is a long silence on the tape. Miss Bridges reached for a cigarette and took a long time to light it. L.M.*]

M.B: I'm sorry . . . I'm being silly. It's only that I haven't spoken about Camilla for years. I've thought of her. Maybe what happened all those years ago has turned me into the kind of writer I am.

L.M: Would you mind telling me about it?

M.B: No, I don't think I'd mind. I've never really told anyone before. I don't know why I've never written about it. Have you got enough tape on this machine? Well then . . . Camilla.

From the very first time I saw her, I wanted passionately to be her best friend. Camilla was beautiful. She had long, dark hair in plaits and wide grey eyes. There was nothing particularly remarkable about her clothes, and yet she always made the rest of us feel like scarecrows. It wasn't just her looks, though. She was peaceful, somewhere deep inside herself, and you felt, you just knew, that being her friend would be the best thing that could ever happen to you.

She wasn't unfriendly. You mustn't think that. She was always very nice to me, and to Sarah and the other big girls, but she would not, whatever I did, commit herself. She would never say, 'I'll be your best friend.' Do you still have best friends nowadays? In my childhood, your best friend was the most important thing in your life. I was quite shameless in my pursuit of Camilla. I used to pass her notes during lessons, but she never answered. She would smile, and nod her head towards Mrs Ashton as if to say: 'Watch out. Teacher will see you.'

Then it was my tenth birthday. I'd persuaded my parents to have a fancy dress party to which all my friends would come. I arrived at school carrying the invitations and handed them out with some pride. At breaktime, sitting on the steps of the school hut with Camilla, I said:

'You will come, won't you? It'll be such fun. It's on Sunday afternoon.'

Camilla shook her head. 'Oh, Monica, I'm sorry. I have to go away on Friday to visit my uncle, who has a rubber plantation up the coast. On Sunday afternoon when you're having your party, we'll be driving back. I wish I could come.'

'Why don't you try?' I said. 'Try asking your parents to leave a bit earlier, so that you can get there for a while, at least.'

Camilla looked at me very hard then. I haven't forgotten her look, even after forty odd years. I shivered then in spite of the sun burning down on me, and I'm shivering now as I tell you. Camilla said: 'I'll try. I'll try and come.'

[*There is a long pause on the tape here. L.M.*]

L.M: Did she come? To your party, I mean?

M.B: Oh, yes, she came. We recognized her in spite of her fancy dress. She came as a dragon, wearing a huge elaborate Chinese dragon-mask over her head, so that unless you knew Camilla very well, you would never guess. We knew it was her by the way she walked, but I don't think anyone else noticed that she was wearing her bracelet. She always wore it on her left wrist. It was an identity disc with her name engraved on it. I welcomed her at the door

and said: 'Oh, Camilla, I'm so glad you came after all. We're having a lovely party.'

And it was a lovely party. I don't think I've ever had a better birthday. The food was delicious, the games were fun and I won a lot of them, all my presents were wonderful, and best of all, better than anything, Camilla said them at last, the words I wanted to hear more than anything. Everyone else had gone off on a treasure hunt round the garden and we were standing on the verandah together.

'Monica, listen,' she said, 'you've asked me and asked me, and I've never said anything, but I would have liked to be your best friend. I really would. I want you to know that, whatever happens. Do you promise?'

'I promise,' I answered, and she said:

'I haven't got a proper present for you, but you can have my bracelet.' She took it off and fastened it round my wrist.

'But, Camilla,' I said, 'you always wear that . . . you can't give it to me. It's got your name on it and everything.'

'I don't need it . . .' Camilla started to say, and then my mother came out and bustled us towards the table where my birthday cake was standing ready to be cut.

The next day, it was school again. I couldn't wait to get there, to take up my new position as Camilla's official best friend. Sarah and I waited behind the frangipani tree for Mrs Ashton, giggling as we thought of the early-morning kiss, and wondering what our teacher would be wearing. Mrs Ashton didn't kiss Brian. It was the first time she had ever left the car without looking fondly behind her.

'They must have had a fight,' Sarah said knowledgeably. 'Her eyes are all red. She's been crying.'

I said nothing. I don't think I knew that grown-ups could cry.

When we had sat down for the first lesson, Mrs Ashton stood up.

'Children,' she said and her voice was thick with tears, 'I have some very sad news for all of you. Camilla won't be coming back . . . there was a terrible accident yesterday afternoon at about 2.00. Camilla and her parents were driving home and . . .' Mrs Ashton swallowed. 'They must have been going very fast. The car went off the road. They were killed at once. I'm sure,' Mrs Ashton added, 'that it happened far too quickly for them to have known anything about it. They can't possibly have suffered.'

She sat down suddenly and put her face in her hands. We all sat and stared at one another. I felt . . . I don't know what I felt except cold, very cold. Then I thought: there must have been a mistake. Camilla was at my party at 3.30 so how could she have died at 2.00? I turned to Sarah. She knew the truth.

'Sarah,' I began.

'Yes?' Tears stood in her eyes.

'Sarah, did you see the person in the Chinese dragon-mask at my party?'

'Yes,' said Sarah. 'I wonder who it can have been. I was sure it was Camilla, but it can't have been, can it?' Sarah burst into tears, and I said nothing. I've never said anything about it to anyone, till now. No one would have believed me.

L.M: Thank you so much for telling me. I'm sorry about your friend. Can you tell me how long ago this was?

M.B: This happened in 1953. I can see by your face that you think writing all these ghost stories has addled my brain. You're being polite, but I don't think you quite believe that Camilla came back from the dead with the gift of her friendship. Am I right?

L.M: No, I do believe you. Truly.

M.B: I think you do . . . isn't that amazing? And you will be rewarded for your belief. I'll show you the proof.

L.M: Is there proof?

M.B: I've never shown anyone before.
[*There's the sound of M.B. opening a locked drawer in her desk. L.M.*]
There. Tell me what you see.

L.M: It's a silver bracelet. There's an identity disc on it, engraved with the name: Camilla.

Dannie Abse

There is the story of a deserted island
where five men walked down to the bay.

The story of this island is
that three men would two men slay.

Three men dug two graves in the sand,
three men stood on the sea-wet rock,
three shadows moved away.

There is the story of a deserted island
where three men walked down to the bay.

The story of this island is
that two men would one man slay.

Two men dug one grave in the sand,
two men stood on the sea-wet rock,
two shadows walked away.

There is the story of a deserted island
where two men walked down to the bay.

The story of this island is
that one man would one man slay.

One man dug one grave in the sand,
one man stood on the sea-wet rock,
one shadow moved away.

There is the story of a deserted island
where four ghosts walked down to the bay.

The story of this island is
that four ghosts would one man slay.

Four ghosts dug one grave in the sand,
four ghosts stood on the sea-wet rock;
five ghosts moved away.

The Shiner

Grace Hallworth

Late one night a policeman was riding home when he saw a young woman walking along a road. The policeman was young and eager for a bit of excitement. He was also a new recruit and did not have the experience which encourages caution, so he approached the woman.

'A young lady like you should not be out alone at this hour,' he said.

The woman appeared to be shy for she kept her face hidden and did not reply.

The policeman persisted. 'Would you like a lift home on my bicycle?' he asked.

The woman continued walking and said nothing.

Her attitude made him more determined. 'Come now, lady,' he urged. 'Surely you don't believe I would harm you. Just tell me where you live and I will see you home safely.'

The woman stopped and shyly said, 'It's not far from here but if you insist I will go with you.'

She was hoisted on the crossbar and they set off pleasantly enough. The policeman was teasing and joking, and though the woman was laughing merrily she hardly said a word. They had not gone very far when the policeman found that he could not turn the pedals as easily as before, yet the road was quite flat. He thought to himself, 'She was as light as a bundle of feathers when we started off. Now she is a dead weight.' He began to feel uneasy as he remembered he had not seen her face.

'Lady, you didn't tell me where you live,' he said.

'Just a little way to go. Keep straight on,' she replied with a strange laugh.

After a few seconds passed with no more directions, he asked, 'Lady, how much further?'

Her reply was the same as before, 'Just a little way to go. Keep straight on.'

The policeman was labouring with the load he carried. He felt as though his lungs were bursting and fear seeped through every pore of his skin as he recalled that there were no houses ahead of them. He was heading for a part of the town where the only buildings were warehouses. Beyond them lay the open sea.

A number of small boats and trawlers were bobbing up and down in the harbour. A light from one of them shone on the silver buttons of the policeman's tunic which formed the pattern of a cross. The emblem of the Star of Bethlehem on his helmet glinted in the dark.

He braked and was about to accuse the woman of making a fool of him when she leapt off the bicycle and turned to face him. As she did so she saw the two emblems. She screamed and hid her head in her hands but not before the policeman had seen eyes like burning coals in a corpse-like face. From her mouth burst flames of fire. He jumped on his bicycle and rode away as fast as his trembling legs could pedal, her shrill fiendish laughter pursuing him as he fled. The echo of her words rang in his head for several days, 'You lucky, you lucky to be wearing that uniform, or your soul would be mine tonight and your body food for the fish. You luckeeeeee. . .'

Since that day the policeman has kept the buttons on his tunic and the badge on his helmet so well polished that he is known as 'The Shiner'.

Prince Kano

Edward Lowbury

In a dark wood Prince Kano lost his way
And searched in vain through the long summer's day.
At last, when night was near, he came in sight
Of a small clearing filled with yellow light,
And there, bending beside his brazier, stood
A charcoal burner wearing a black hood.
The Prince cried out for joy: 'Good friend, I'll give
What you will ask: guide me to where I live.'
The man pulled back his hood: he had no face—
Where it should be there was an empty space.

Half dead with fear the Prince staggered away,
Rushed blindly through the wood till break of day;
And then he saw a larger clearing, filled
With houses, people; but his soul was chilled.
He looked around for comfort, and his search
Led him inside a small, half-empty church
Where monks prayed. 'Father,' to one he said,
'I've seen a dreadful thing; I am afraid.'
'What did you see, my son?' 'I saw a man
Whose face was like . . .' and, as the Prince began,
The monk drew back his hood and seemed to hiss,
Pointing to where his face should be, 'Like this?'

Wiley and the Hairy Man

Virginia Haviland

Wiley's pappy was a bad man and no-count. He stole watermelons in the dark of the moon. He was lazy, too, and slept while the weeds grew higher than the cotton. Worse still, he killed three martins and never even chunked at a crow.

One day he fell off the ferry boat where the river is quicker than anywhere else and no one ever found him. They looked for him a long way down river and in the still pools between the sand-banks, but they never found him. They heard a big man laughing across the river, and everybody said, 'That's the Hairy Man.' So they stopped looking.

'Wiley,' his mammy told him, 'the Hairy Man's got your pappy and he's goin' to get you if you don't look out.'

'Yas'm,' he said. 'I'll look out. I'll take my hound-dogs everywhere I go. The Hairy Man can't stand no hound-dog.'

Wiley knew that because his mammy had told him. She knew because she came from the swamps by the Tombigbee River and knew conjure magic.

One day Wiley took his axe and went down in the swamp to cut some poles for a hen-roost and his hounds went with him. But they took out after a young pig and ran it so far off Wiley couldn't even hear them yelp.

'Well,' he said, 'I hope the Hairy Man ain't nowhere round here now.'

He picked up his axe to start cutting poles, but he looked up and there came the Hairy Man through the trees grinning. He was sure ugly and his grin didn't help much. He was hairy all over. His eyes burned like fire and spit drooled all over his big teeth.

'Don't look at me like that,' said Wiley, but the Hairy Man kept coming and grinning, so Wiley threw down his axe and climbed up a big bay tree. He saw the Hairy Man didn't have

feet like a man but like a cow, and Wiley never had seen a cow up a bay tree.

'What for you done climb up there?' the Hairy Man asked Wiley when he got to the bottom of the tree.

Wiley climbed nearly to the top of the tree and looked down. Then he climbed plumb to the top.

'How come you climbin' trees?' the Hairy Man said.

'My mammy done tole me to stay away from you. What you got in that big sack?'

'I ain't got nothin' yet.'

'Gwan away from here,' said Wiley, hoping the tree would grow some more.

'Ha,' said the Hairy Man and picked up Wiley's axe. He swung it about and the chips flew. Wiley grabbed the tree close, rubbed his belly on it and hollered, 'Fly, chips, fly, back in your same old place.'

The chips flew and the Hairy Man cussed and damned. Then he swung the axe and Wiley knew he'd have to holler fast. They went to it tooth and toe-nail then, Wiley hollering and the Hairy Man chopping. He hollered till he was hoarse and he saw the Hairy Man was gaining on him.

'I'll come down part of the way,' he said, 'if you'll make this bay tree twice as big around.'

'I ain't studyin' you,' said the Hairy Man, swinging the axe.

'I bet you can't,' said Wiley.

'I ain't going to try,' said the Hairy Man.

Then they went to it again, Wiley hollering and the Hairy Man chopping. Wiley had about yelled himself out when he heard his hound-dogs yelping way off.

'Hyeaaah, dog,' hollered Wiley, and they both heard the hound-dogs yelping and coming jam-up. The Hairy Man looked worried.

'Come on down,' he said, 'and I'll teach you conjure.'

'I can learn all the conjure I want from my mammy.'

The Hairy Man cussed some more, but he threw the axe down and took off through the swamp.

When Wiley got home he told his mammy that the Hairy Man had most got him, but his dogs ran him off.

'Did he have his sack?'

'Yas'm.'

'Next time he come after you, don't you climb no bay tree.'

'I ain't,' said Wiley. 'They ain't big enough around.'

'Don't climb no kind o' tree. Just stay on the ground and say "Hello, Hairy Man." You hear me, Wiley?'

'No'm.'

'He ain't goin' to hurt you, child. You can put the Hairy Man in the dirt when I tell you how to do him.'

'I puts him in the dirt and he puts me in that big sack. I ain't puttin' no Hairy Man in the dirt.'

'You just do like I say. You say, "Hello, Hairy Man." He says, "Hello, Wiley." You say, "Hairy Man, I done heard you about the best conjureman 'round here." "I reckon I am." You say, "I bet you cain't turn yourself into no giraffe." You keep tellin' him he cain't and he will. Then you say, "I bet you cain't turn yourself into no 'possum." Then he will, and you grab him and throw him in the sack.'

'It don't sound just right somehow,' said Wiley, 'but I will.' So he tied up his dogs so they wouldn't scare away the Hairy Man, and went down to the swamp again. He hadn't been there long when he looked up and there came the Hairy Man grinning through the trees, hairy all over and his big teeth showing more than ever. He knew Wiley came off without his hound-dogs. Wiley nearly climbed a tree when he saw the big sack, but he didn't.

'Hello, Hairy Man,' he said.

'Hello, Wiley.' He took the sack off his shoulder and started opening it up.

'Hairy Man, I done heard you are about the best conjureman round here.'

'I reckon I is.'

'I bet you cain't turn yourself into no giraffe.'

'Shucks, that ain't no trouble,' said the Hairy Man.

'I bet you cain't do it.'

So the Hairy Man twisted round and turned himself into a giraffe.

'I bet you cain't turn yourself into no alligator,' said Wiley.

The giraffe twisted around and turned into an alligator, all the time watching Wiley to see he didn't try to run.

'Anybody can turn theyself into something big as a man,' said Wiley, 'but I bet you cain't turn yourself into no 'possum.'

The alligator twisted around and turned into a 'possum, and Wiley grabbed it and threw it in the sack.

Wiley tied the sack up as tight as he could and then he threw it in the river. He started home through the swamp and he looked up and there came the Hairy Man grinning through the trees. Wiley had to scramble up the nearest tree.

The Hairy Man gloated: 'I turned myself into the wind and blew out. Wiley, I'm going to set right here till you get hungry and fall out of that bay tree. You want me to learn you some more conjure?'

Wiley studied a while. He studied about the Hairy Man and he studied about his hound-

dogs tied up most a mile away.

'Well,' he said, 'you done some pretty smart tricks. But I bet you cain't make things disappear and go where nobody knows.'

'Huh, that's what I'm good at. Look at that old bird-nest on the limb. Now look. It's done gone.'

'How I know it was there in the first place? I bet you cain't make something I know is there disappear.'

'Ha ha!' said the Hairy Man. 'Look at your shirt.'

Wiley looked down and his shirt was gone, but he didn't care, because that was just what he wanted the Hairy Man to do.

'That was just a plain old shirt,' he said. 'But this rope I got tied round my breeches has been conjured. I bet you cain't make it disappear.'

'Huh, I can make all the rope in this county disappear.'

'Ha ha ha,' said Wiley.

The Hairy Man looked mad and threw his chest way out. He opened his mouth wide and hollered loud.

'From now on all the rope in this county has done disappeared.'

Wiley reared back, holding his breeches with one hand and a tree-limb with the other.

'Hyeaaah, dog,' he hollered loud enough to be heard more than a mile off.

When Wiley and his dogs got back home his mammy asked him did he put the Hairy Man in the sack.

'Yes'm, but he done turned himself into the wind and blew right through that old sack.'

'That *is* bad,' said his mammy. 'But you done fool him twice. If you fool him again he'll leave you alone. He'll be mighty hard to fool the third time.'

'We got to study up a way to fool him, mammy.'

'I'll study up a way tereckly,' she said, and sat down by the fire and held her chin between her hands and studied real hard. But Wiley wasn't studying anything except how to keep the Hairy Man away. He took his hound-dogs out and tied one at the back door and one at the front door. Then he crossed a broom and an axe-handle over the window and built a fire in the fire-place. Feeling a lot safer, he sat down and helped his mammy study. After a little while his mammy said, 'Wiley, you go down to the pen and get that little suckin' pig away from that old sow.'

Wiley went down and snatched the sucking pig through the rails and left the sow grunting and heaving in the pen. He took the pig back to his mammy and she put it in his bed.

'Now, Wiley,' she said, 'you go on up to the loft and hide.'

So he did. Before long he heard the wind howling and the trees shaking, and then his dogs started growling. He looked out through a knot-hole in the planks and saw the dog at the front door looking down toward the swamps, with his hair standing up and his lips drawn back in a snarl. Then an animal as big as a mule with horns on its head ran out of the swamp past the house. The dog jerked and jumped, but he couldn't get loose. Then an animal bigger than a great big dog with a long nose and big teeth ran out of the swamp and growled at the cabin. This time the dog broke loose and took after the big animal, who ran back down into the swamp. Wiley looked out another chink at the back end of the loft just in time to see his other dog jerk loose and take out after an animal which might have been a 'possum, but wasn't.

'Law-dee,' said Wiley. 'The Hairy Man is coming here, sure.'

He didn't have long to wait, because soon enough he heard something with feet like a cow scrambling around on the roof. He knew it was the Hairy Man, because he heard him swear when he touched the hot chimney. The Hairy Man jumped off the roof when he found out there was a fire in the fire-place and came up and knocked on the front door as big as you please.

'Mammy,' he hollered, 'I done come after your baby.'

'You ain't going to get him,' mammy hollered back.

'Give him here or I'll set your house on fire with lightning.'

'I got plenty of sweet-milk to put it out with.'

'Give him here or I'll dry up your spring, make your cow go dry, and send a million boll-weevils out of the ground to eat up your cotton.'

'Hairy Man, you wouldn't do all that. That's mighty mean.'

'I'm a mighty mean man. I ain't never seen a man as mean as I am.'

'If I give you my baby will you go on way from here and

leave everything else alone?'

'I swear that's just what I'll do,' said the Hairy Man, so mammy opened the door and let him in.

'He's over there in that bed,' she said.

The Hairy Man came in grinning like he was meaner than he said. He walked over to the bed and snatched the covers back.

'Hey,' he hollered, 'there ain't nothing in this bed but a old suckin' pig.'

'I ain't said what kind of baby I was giving you, and that suckin' pig sure belong to me before I gave it to you.'

The Hairy Man raged and yelled. He stomped all over the house gnashing his teeth. Then he grabbed up the pig and tore out through the swamp, knocking down trees right and left. The next morning the swamp had a wide path like a cyclone had cut through it, with trees torn loose at the roots and lying on the ground. When the Hairy Man was gone Wiley came down from the loft.

'Is he done gone, mammy?'

'Yes, child. That old Hairy Man cain't ever hurt you again. We done fool him three times.'

THE START OF A MEMORABLE HOLIDAY

Roy Fuller

Good evening, sir. Good evening, ma'am. Good evening, little ladies.
From all the staff, a hearty welcome to the Hotel Hades.
Oh yes, sir, since you booked your rooms we have been taken over
And changed our name—but for the better—as you'll soon discover.
Porter, Room 99! Don't worry, sir—just now he took
Much bulkier things than bags on his pathetic iron hook.
The other room, the children's room? I'm very pleased to say
We've put them in the annexe, half a mile across the way.
They'll have a nearer view there of the bats' intriguing flying,
And you, dear sir and madam, won't be troubled by their crying
—Although I'm sure that neither of them's frightened of the gloom.
Besides, the maid will try to find a candle for their room.
Of course, ma'am, we've a maid there, she's the porter's (seventh) wife:
She'll care for these dear children quite as well as her own life.
The journey's tired them? Ah, tonight they won't be counting sheep!
I'll see they have a nice hot drink before they're put to sleep.
Don't be too late yourselves, sir, for the hotel's evening meal:
I hope that on the menu will be some roast milk-fed veal.
If you'll forgive me, I must stoke the ovens right away:
It's going to be (excuse the joke) hell in this place today!
Yes, I do all the cooking and the getting of the meat:
Though we're so far from the shops we've usually something fresh to eat.
Of course, it isn't always veal, and when the school terms start
Joints may get tougher. But our gravy still stays full of heart!

The Great Swallowing Monster

Geraldine McCaughrean

Tiko was a coward. He knew it. He had always known it. He was just as much a coward as his sister Marra, and she said every day how she wished she could be brave.

Oh, there was that time beside the sheep fold. A hyena came sniffing out of the dawn, meaning to steal a lamb. And Tiko and Marra threw stones and scared the hyena away. Their mother and father said how brave they were to do it. But Tiko knew he was really a coward. At night he dreamed of the Great Swallowing Monster and woke up too terrified to go to sleep again. No, Tiko was a coward. There was no escaping that.

Oh yes, there was that time among the cows. A yellow lioness came stalking through the yellow grass, her round ears flickering like the leaves, and the smell of beef in her velvet nose. And Tiko and Marra snatched up sticks and shrieked and whooped till the cows began to run. The lioness went home hungry. The people in the village said how brave the children were to do it. But Tiko knew he was really a coward. He only had to think about the Great Swallowing Monster and his mouth felt dry and his palms felt wet and his heart raced. Tiko was a coward. There was no doubt about it. It made him very ashamed.

Oh, and there was that time when a snake came slithering out of the dark and into the middle of the village—into the long hut where everybody slept, up to the cradle of the chief's newborn baby. And Tiko snatched up a broom and Marra a pestle, and they swept the wriggling, hissing snake out-of-doors and told it, 'Don't come back!' The chief said they were very brave to do it. But Tiko only shook his head and sighed. He was thinking about the Great Swallowing Monster, wondering, 'Is it out there now? Is it coming this way? Will it be here tomorrow?' He was a coward, no question about it.

Whenever the tribe gathered together for a story, the old Storyteller would talk of the Great Swallowing Monster. 'He's as huge as a hill. As hungry as a hole. As wicked as Evil itself!' He told how the Great Swallowing Monster ate up whole villages, ate up whole tribes, ate up herds of cows and clumps of trees. And whenever the Storyteller told such stories, Tiko crept to bed afterwards trembling with fright, and dreamed bad dreams and woke up crying.

One night Tiko whispered through the dark to his sister, 'I dreamed I heard it coming! I dreamed I heard its footsteps!'

'You didn't dream it,' said Marra.

She was peeping out through the door of the hut. And there, coming closer and closer with every breath, every heartbeat, was the Great Swallowing Monster. It blocked out the moon and half of the stars. Running on its two legs, like an ostrich, it shook the ground with its great splay feet. When it splashed through the river, water flew up to this side and that, and left the river-bed dry. When the Monster opened its mouth, a flock of geese flew inside and disappeared. Its long neck snaked, and its head seemed to butt at the planets and gobble up the stars.

Tiko ran to his mother and shook her by the hair, 'Mother! Mother, get up! The Great Swallowing Monster is coming! We must all run away!'

'Hush now,' she said. 'It's another one of your dreams. Go back to bed.'

And all the while, the Monster got closer.

Marra ran to the Chief and shook him by the beard. 'Chief! Chief, get up! The Great Swallowing Monster is coming! We must all run away!'

'Less noise! People are trying to sleep. Tell me about it in the morning. No need to be afraid: we have warriors to protect us from danger, don't we? Go back to bed and leave worrying to the grown-ups.'

And all the while the Monster got closer.

So Tiko and Marra ran to the Storyteller and shook him by both ears. 'Sir! Sir, get up! The Great Swallowing Monster is coming! We must all run away!'

'Don't be so foolish,' snorted the Storyteller. 'The Great Swallowing Monster is only a legend. It's only something spoken of in stories. And I thought you were such big brave children! Go back to bed.'

And all the while, the Monster got closer.

So Tiko and Marra ran out into the moonlight. They ran between the very legs of the Great Swallowing Monster just as it opened its mouth to swallow the hut.

'*Grrmmschschlrpsmckglp!*' said the Great Swallowing Monster and swallowed down the long-hut and everyone in it: swallowed

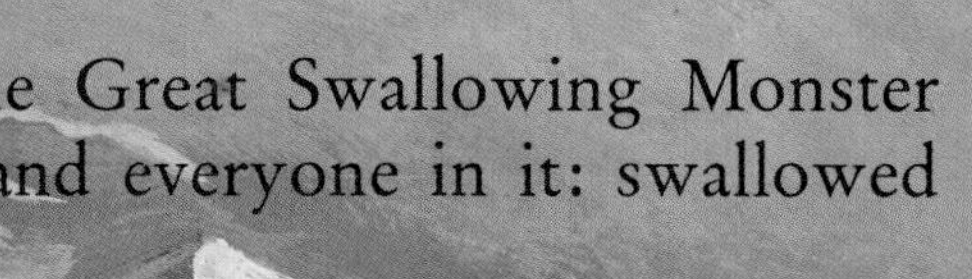

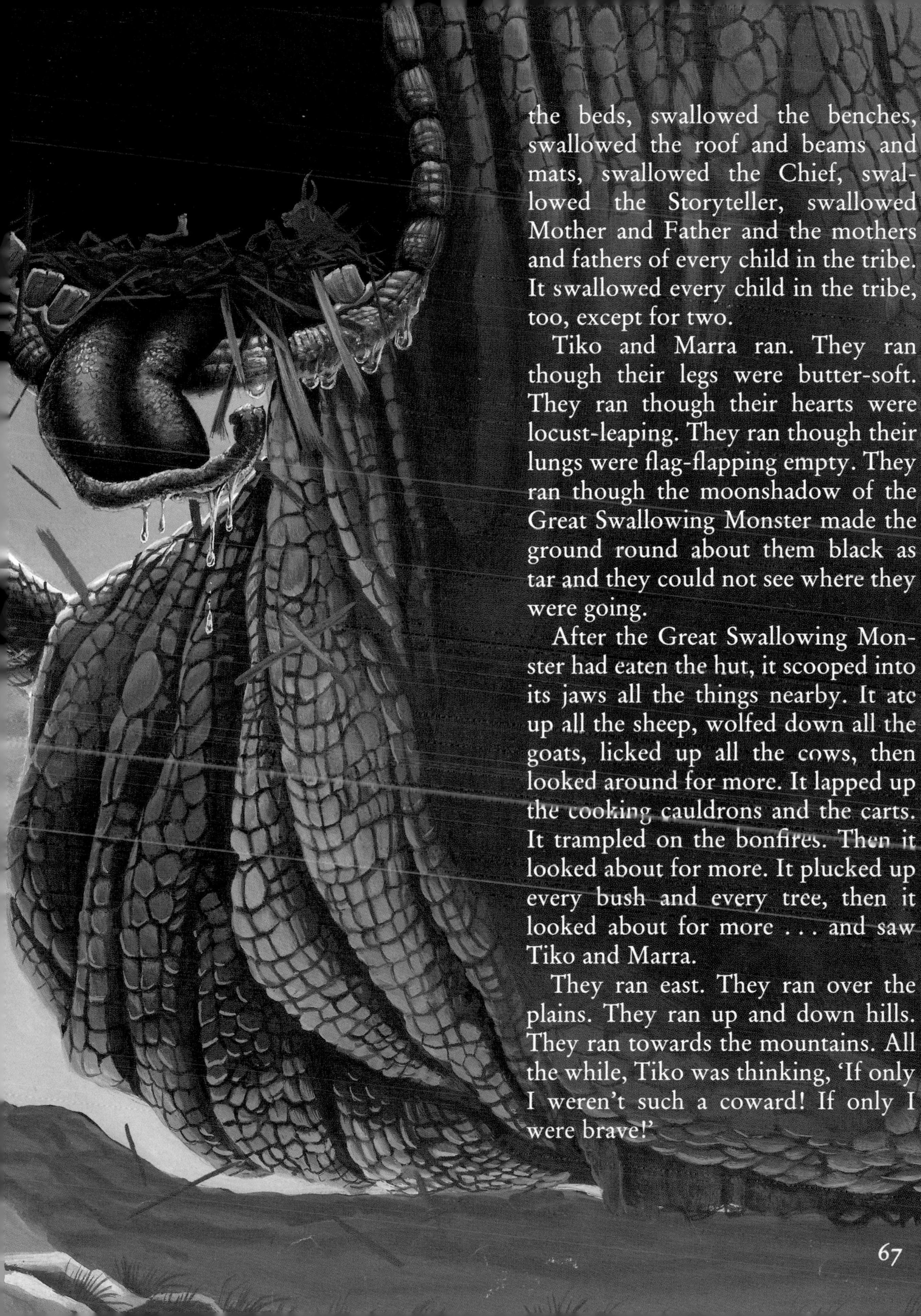

the beds, swallowed the benches, swallowed the roof and beams and mats, swallowed the Chief, swallowed the Storyteller, swallowed Mother and Father and the mothers and fathers of every child in the tribe. It swallowed every child in the tribe, too, except for two.

Tiko and Marra ran. They ran though their legs were butter-soft. They ran though their hearts were locust-leaping. They ran though their lungs were flag-flapping empty. They ran though the moonshadow of the Great Swallowing Monster made the ground round about them black as tar and they could not see where they were going.

After the Great Swallowing Monster had eaten the hut, it scooped into its jaws all the things nearby. It ate up all the sheep, wolfed down all the goats, licked up all the cows, then looked around for more. It lapped up the cooking cauldrons and the carts. It trampled on the bonfires. Then it looked about for more. It plucked up every bush and every tree, then it looked about for more . . . and saw Tiko and Marra.

They ran east. They ran over the plains. They ran up and down hills. They ran towards the mountains. All the while, Tiko was thinking, 'If only I weren't such a coward! If only I were brave!'

Beside him Marra gasped as she ran, 'I see now! I see why we came this way! You're so clever and so brave, Tiko!'

'You do?' said Tiko. 'I am?' He looked back over his shoulder.

The Great Swallowing Monster was coming closer. It would have caught them long before, but it stopped here to drink up a lake and there to swallow down a herd of wildebeest; now to snatch up a flock of rhea-birds, and then to chew on a hill.

On and on ran Tiko and Marra, down a valley and between two mountains. And behind them came the Great Swallowing Monster, its baggy brown body bulging bigger and bigger as it ate up everything in sight. It was right behind them; its little eye swooped closer; its jaws parted; its tongue rolled out longer and red.

'*Ooooffphphph!*'

Tiko looked back. The huge, hurtling hulk of the Great Swallowing Monster was still behind him. Its neck was still stretched out towards him. Its mouth was still gaping to eat him up. But the Great Swallowing Monster's leathery brown bag of a body was so full of food that it had jammed between the two mountains. It could not go forward and it could not go back.

It writhed and wriggled. But it was wedged between the mountains as surely as May is wedged between April and June. It stretched and struggled. But it was jammed between the mountains as surely as Tuesday is jammed between Monday and Wednesday.

Tiko and Marra sat down and rested. 'How clever of you, brother, to trap the Great Swallowing Monster like that,' said Marra. But Tiko said nothing.

Soon they got up and went back the way they had come. They walked up to the Monster. They climbed up over the Monster's lip. They scrambled up the Monster's teeth and swung up on to the Monster's nose. Then they brushed the dust off their hands—right in at the nostrils of the Great Swallowing Monster.

As big as caves, those nostrils were. When the Monster sneezed, the sound shook the mountains and made seven rivers change course.

Out flew the lake in a rainbow of spray. Out flew the rhea-birds, all wet from swimming in the lake. Out flew the cows and goats and sheep, and the wildebeest. Out flew the cauldrons and carts and

trees. Out flew the Storyteller and the Chief and a great many other people. Then best of all, out flew their mother and father, though they were angry because of the long walk home.

'How wonderful you were, Tiko!' said everyone.

'How clever!'

'How brave! How brave! How brave!'

Tiko took a deep breath. 'I wasn't brave at all. I was just scared.'

And Marra said, 'Me too.'

The Chief looked at the Storyteller. The Storyteller looked at Mother and Mother looked at the Chief. Then the Chief looked at Tiko and Marra and threw his hands in the air. 'Well *of course* you were scared! What fool isn't afraid of a Monster as big as a city? But you came back, didn't you? You came back to rescue us!'

Tiko and Marra were puzzled. 'Well, we had to, didn't we? There was nobody else to do it.'

It did not matter how much Tiko and Marra tried to explain. They were carried home shoulder-high, and given garlands of flowers, because they were so brave.

'Nobody realizes what a coward I really am,' said Tiko.

'Nor me,' said Marra and sighed.

And now when they look towards the East and see three mountains instead of two, they still tremble a little.

But then so does the Chief.

The Slitherydee

Robert Scott

THE SLITHERYDEE
SLIPPED OUT OF THE SEA
IT CAUGHT ALL THE OTHERS
– BUT IT DIDN'T CATCH ME

THE SLITHERYDEE
SLIPPED OUT OF THE SEA
IT ATE ALL THE OTHERS

– BUT IT DIDN'T –
SLURP....

Who's that?

James Kirkup

Who's that
stopping at
my door in the
dark, deep
in the dead of the moonless night?

Who's
that in the quiet
blackness,
darker than dark?

Who
turns the han-
dle of my door, who
turns the old brass hand-
le of
my door with never a sound, the handle
that always
creaks and rattles and
squeaks but
now
turns
without a sound, slowly
slowly
slowly
round?

Who's that moving through the floor
as if it were a lake, an open door? Who
is it who passes through
what can never be passed through,
who passes through
the rocking-chair
without rocking it,
who passes through
the table without knocking it, who
walks out of the cupboard without unlocking it?
Who's that? Who plays with my toys
with no noise, no
noise?

Who's that? Who is it
silent and silver
as things in mirrors, who's
as slow as feathers,
shy as the shivers,
light as a fly?

Who's that who's that
as close as
close as a hug, a kiss—

Who's THIS?

Bush Lion

Robert Scott

The way through the bush was long and weary. He had met with a Bush woman carrying a small child on her back, so they travelled together. Now they had run out of food. Hot and tired, they squatted beneath a tree.

'I think there are cattle over there,' he said, pointing through the shimmering heat.

'I see them.'

'We could do with meat.'

'Yes.'

'If we don't eat we won't finish our journey at all.'

'No.'

He looked sideways at the Bush woman. 'If what I've heard about you people is true,' he said, 'you can turn yourself into a lion.'

She didn't answer.

'And then you could catch one of those animals and we could both eat.' He smiled at her. 'That's what they say,' he said.

'Do you believe what they say?'

He shook his head. 'But we must have food.'

'If I changed into a lion you would be frightened.'

He looked out at the grazing animals. 'The only thing I'm afraid of at the moment is dying of hunger,' he said.

'And you wouldn't trust me on the rest of the journey.'

He turned to answer her, but saw her nails become long and sharp, her face grow furry, her eyes golden and savage. She dropped her baby and tore off her wrap. A low snarl rumbled in her throat. She stretched her jaws and glared at the man who was, by now,

halfway up the tree. She dropped to her belly and disappeared silently in the direction of the grazing animals.

From the safety of his tree he saw the animals break and stampede, heard the squeal of her victim and her roar of triumph. Before long she reappeared dragging a young eland, which she laid beneath the tree. She looked up at the man and gave a loud, coughing roar.

'All right!' he cried. 'All right, I believe you. Now suppose you change back again.'

The lion growled and lashed her tail.

'So now I have to die of hunger up a tree instead of on the ground?'

She settled by the dead eland and looked up at him, growling.

'*Please* change back,' he said.

The lion rose and stretched. She reared herself against the trunk of the tree, tearing the bark with her claws as the man scrambled higher in the branches. Then, tired of her teasing, she sat with her back to him and the man saw the Bush woman slowly emerge and the lion disappear.

'Why don't you come down?' she said.

He made no move.

'It's quite safe. There's nobody here but me. No animal either, except', she nodded at the foot of the tree, 'we have food now.'

It took him longer to get down than it had to climb the tree. By the time he reached the ground the Bush woman had dressed herself and was nursing her baby. He stood close to the tree.

She smiled at him. 'Why don't we eat?' she said.

'Yes.'

'And we can carry some with us,' she said. 'In case we get hungry again.'

The man didn't answer.

'You don't want to believe everything you hear,' the Bush woman told him. 'Or see.'

Eyes of the Cat

Ruskin Bond

Her eyes seemed flecked with gold when the sun was on them. And as the sun set over the mountains, drawing a deep red wound across the sky, there was more than gold in Binya's eyes, there was anger; for she had been cut to the quick by some remarks her teacher had made—the culmination of weeks of insults and taunts.

Binya was poorer than most of the girls in her class and could not afford the tuitions that had become almost obligatory if one was to pass and be promoted. 'You'll have to spend another year in the 9th,' said Madam. 'And if you don't like that, you can find another school—a school where it won't matter if your blouse is torn and your tunic is old and your shoes are falling apart.'

Madam had shown her large teeth in what was supposed to be a good-natured smile, and all the girls had tittered dutifully. Sycophancy had become part of the curriculum, in Madam's private academy for girls.

On the way home in the gathering gloom, Binya's two companions commiserated with her.

'She's a mean old thing,' said Usha. 'She doesn't care for anyone but herself.'

'Her laugh reminds me of a donkey braying,' said Sunita, who was more forthright.

But Binya wasn't really listening. Her eyes were fixed on some point in the far distance, where the pines stood in silhouette against a night sky that was growing brighter every moment. The moon was rising, a full moon, a moon that meant something very special to Binya, that made her blood tingle and her skin prickle and her hair glow and send out sparks. Her steps seemed to grow lighter,

her limbs more sinewy as she moved gracefully, softly over the mountain path.

Abruptly she left her companions at a fork in the road.

'I'm taking the short cut through the forest,' she said.

Her friends were used to her sudden whims. They knew she was not afraid of being alone in the dark. But Binya's moods made them feel a little nervous, and now, holding hands, they hurried home along the open road.

The short cut took Binya through the dark oak forest. The crooked, tormented branches of the oaks threw twisted shadows across the path. A jackal howled at the moon; a nightjar called from the bushes. Binya walked fast, not out of fear but from urgency, and her breath came in short sharp gasps. Bright moonlight bathed the hillside when she reached her home on the outskirts of the village.

Refusing her dinner, she went straight to her small room and flung the window open. Moonbeams crept over the window-sill and over her arms which were already covered with golden hair. Her strong nails had shredded the rotten wood of the window-sill.

Tail swishing and ears pricked, the tawny leopard came swiftly out of the window, crossed the open field behind the house, and melted into the shadows.

A little later it padded silently through the forest.

Although the moon shone brightly on the tin-roofed town, the leopard knew where the shadows were deepest and merged beautifully with them. An occasional intake of breath, which resulted in a short rasping cough, was the only sound it made.

Madam was returning from dinner at a ladies' club, called the Kitten Club as a sort of foil to their husbands' club affiliations. There were still a few people in the street, and while no one could help noticing Madam, who had the contours of a steam-roller, none saw or heard the predator who had slipped down a side alley and reached the steps of the teacher's house. It sat there silently, waiting with all the patience of an obedient schoolgirl.

When Madam saw the leopard on her steps, she dropped her handbag and opened her mouth to scream; but her voice would not materialize. Nor would her tongue ever be used again, either to savour chicken biryani or to pour scorn upon her pupils, for the leopard had sprung at her throat, broken her neck, and dragged her into the bushes.

In the morning, when Usha and Sunita set out for school, they stopped as usual at Binya's cottage and called out to her.

Binya was sitting in the sun, combing her long black hair.

'Aren't you coming to school today, Binya?' asked the girls.

'No, I won't bother to go today,' said Binya. She felt lazy, but pleased with herself, like a contented cat.

'Madam won't be pleased,' said Usha. 'Shall we tell her you're sick?'

'It won't be necessary,' said Binya, and gave them one of her mysterious smiles. 'I'm sure it's going to be a holiday.'

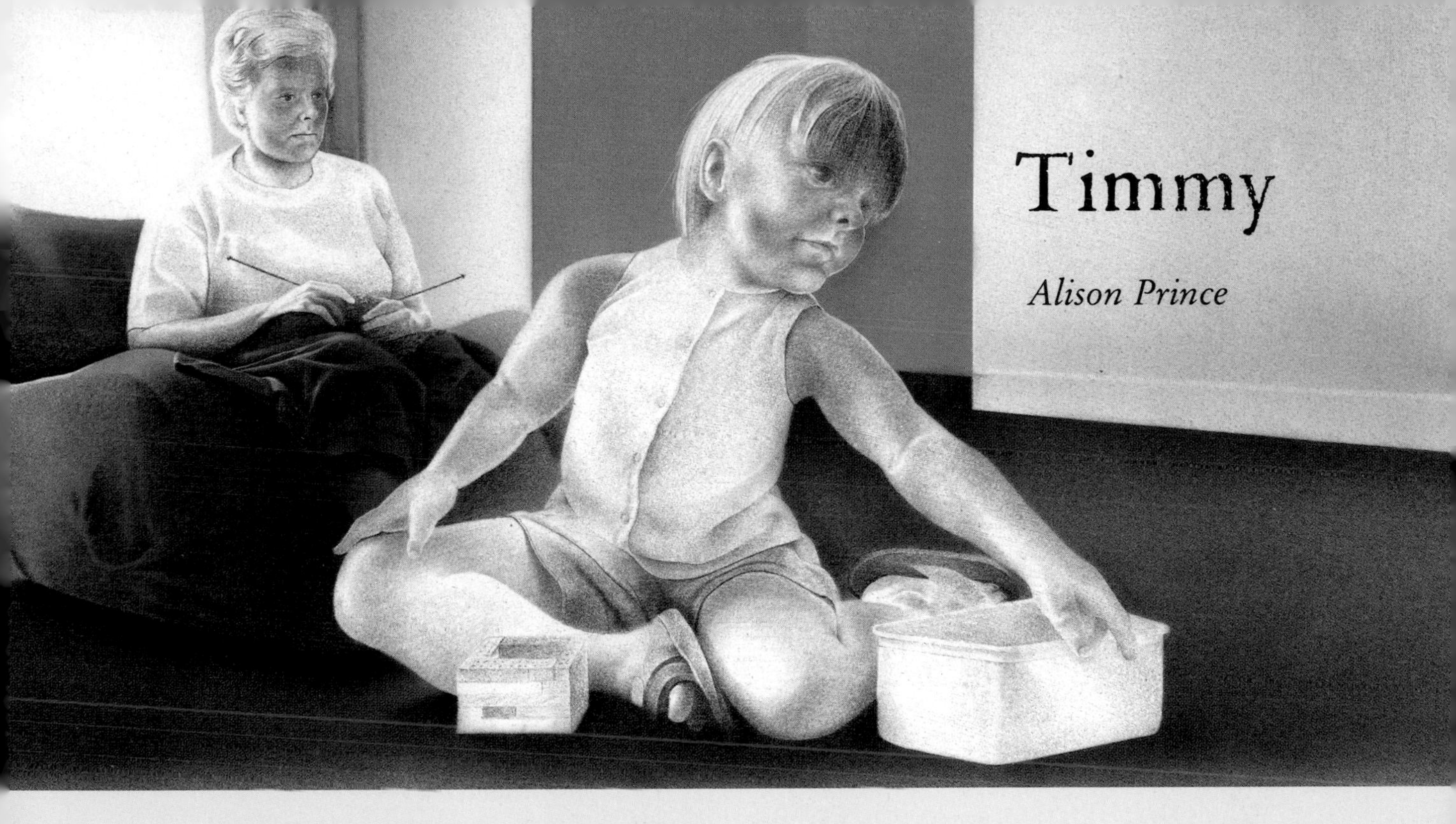

Timmy

Alison Prince

'Timmy,' said Anthony crossly, 'don't take all the long bricks. I want to make a bridge.' His mother let her knitting fall into her lap and stared at him. 'You can't take all the best ones,' Anthony went on, 'just because you're older than me.'

'Anthony,' said his mother. '*Anthony*. Look at me.' Anthony glanced up from his game on the floor. 'You can see me, can't you?' demanded his mother.

'Yes,' said Anthony. He knew what she was going to say next.

'But you can't see Timmy.'

'No,' agreed Anthony.

'Then he isn't *there*,' insisted his mother with a note of desperation.

'He is really,' Anthony said gently. 'Only you can't see him.'

'But neither can you,' his mother pointed out.

Anthony thought. Then he said, 'I can, sort of.'

His mother sighed crossly and took up her knitting again. It was a shawl for the new baby and it had lots of complicated holes and knotty bits in it.

Anthony played with his bricks quietly for a while, trying to pretend that Timmy wasn't there. He usually tried not to talk to Timmy when there were grown-ups about, because it seemed to make them cross, but it was difficult to remember all the time. His mother was the worst of all. She didn't like Timmy a bit.

Soon Anthony knew that Timmy was getting bored. It was no fun when they weren't allowed to talk to each other. Timmy pointed out a colouring-book lying on the floor and suggested that they should colour one of the pictures. 'All right,' Anthony agreed. He picked up a red felt-tip pen and offered Timmy a green one. 'I'll do the engine and you can do the trees.'

Anthony's mother jumped to her feet with a little gasp and said, 'It's time for tea.' She bundled her knitting away, then knelt down on the floor with a grunt of difficulty and then started to scoop the building-bricks into their box. Anthony didn't feel very hungry and Timmy said it wasn't tea-time at all. Timmy didn't like mealtimes much and usually stayed away then.

The sardines on toast smelt very nice and Anthony found that he was quite hungry after all. Timmy had gone off in a huff.

Sipping her cup of tea, Anthony's mother asked, 'How old is your friend?'

'Timmy's five,' said Anthony. 'Can I have a bun?'

'Two years older than you,' said his mother broodingly. She passed Anthony the plate and watched him as he pulled the iced bun out of its paper case and bit the cherry off the top. 'Quite soon,' she said, 'you'll have a real little brother or sister to play with.'

'I know,' said Anthony. 'You said.'

'Well—somebody real is much better than a friend who's not really there, don't you think?'

'But Timmy *is* there,' said Anthony, unruffled.

After a few minutes his mother tried again. 'Why do you call him Timmy?' she asked. 'Has anyone mentioned—told you about someone called Timmy? Someone real?'

'I said, "What's your name?"' said Anthony, 'and he said, "Timmy".'

'Oh.' His mother looked no happier.

'Can I go in the garden?' asked Anthony. 'I've finished my tea.' And Timmy, he noticed, was looking in through the kitchen window.

'Yes,' said his mother with a sigh. She wiped Anthony's sticky fingers with the blue cloth that hung over the edge of the sink and he ran out, rubbing his hands on the front of his T-shirt.

The next day something terrible happened and Anthony lay face down on his bed and wept. His mother found him and gathered him on to her lap. 'Anthony,' she said anxiously, 'whatever is it? Tell Mummy.'

Anthony could not speak for some minutes but when his sobbing abated a little he blurted out, 'Timmy's gone away.'

'Really?' His mother did not sound very sorry. 'Why?'

'I don't know,' said Anthony, sniffing. 'He said he had to go somewhere else.'

'Never mind, darling,' said his mother, hugging him. 'When our new baby's born—'

'He said he'd see me again,' interrupted Anthony, 'but he'd be different, he said.'

'How different?'

'I don't know.' Anthony began to cry again. 'I think he's beastly.'

He was unhappy all the rest of the day, fidgeting round the house and staring out of the window, and when his mother put him to bed he did not want a story.

When Anthony woke the next morning his granny was smiling down at him. 'I've got a lovely surprise for you!' she said. 'You've got a little brother!'

'Oh,' said Anthony. He looked round for Timmy, who was usually sitting on the edge of his bed, and then remembered that he had gone away. He turned on his side and pulled the sheet up over his ears.

'Come on, sleepy-head,' said his granny, shaking Anthony's shoulder gently. 'We can go and see Mummy today, and your new little brother. You want to be up and dressed.'

Reluctantly, Anthony allowed himself to be dressed and washed and given breakfast. Today was just as bad as yesterday. He kept looking for Timmy everywhere although he knew he would not find him, and the house seemed achingly empty.

In the afternoon they took a bus and went to the hospital where Anthony's mother was. They walked along funny-smelling corridors with tiled floors until they came to swing doors with round windows in them. There were two long rows of beds inside, and each bed had a lady lying in it and a little cot beside it. In one of these beds was Anthony's mother. She smiled happily when she saw him coming, and held out her arms to Anthony—but he hardly took any notice because Timmy was there.

He ran to the cot and looked in. 'Timmy!' he said. 'What are you doing there?' He stared in horror at the small, crumpled red face and the tiny fist with a plastic band round the wrist, then ran to his mother. 'What have you done to him?' he shouted. 'Timmy's different!'

Anthony's mother began to cry and his granny kept saying, 'This isn't Timmy dear, this is a new little baby. This isn't Timmy. This is Matthew.' But Anthony could not bear to look into the cot again, where Timmy lay, so dreadfully changed into a helpless, ugly thing. They did not stay very long in the hospital because Anthony's mother kept crying and his granny seemed upset, too.

It was not until they were going home in the bus that Anthony remembered what Timmy had said about seeing him again and being different. This must be what he meant. So perhaps it wasn't his mother's fault. He thought about it for a while, then asked his granny, 'Can babies play with bricks?'

'No,' said his granny. She did not seem very cheerful since they had been in the hospital. After a few moments she added, 'They can when they get bigger. Quite soon.'

Anthony nodded and said no more.

That evening, when she had bathed him and put him to bed, Anthony's granny sat down on the bedside chair and said, 'Anthony, listen. I'm going to tell you something. And I want you to be a very grown-up boy and try to understand. Your Mummy is getting very upset, you see. It's about your friend.'

'Timmy?' asked Anthony, interested.

'That's right. Now, your Mummy doesn't mind you having a friend that nobody can see. That's quite all right. What worries her is that you call him Timmy.'

Anthony tried to say something but his granny went on firmly. 'Just listen for a minute. This is what I want to tell you. Two years before you were born, your Mummy had another baby. If he had lived, he would have been your elder brother. But sometimes babies are not very strong or there is something wrong with them and—'

'I know,' interrupted Anthony. 'He was called Timmy, and he died.'

His grandmother stared at him. Her mouth opened a little and she took her glasses off and rubbed her eyes, then put her glasses on again. 'How do you know?' she demanded almost angrily. 'He was born hundreds of miles away—your Mummy and Daddy moved down here a few months later because they wanted to get right away from where it happened. Nobody knows about it here. You were called Anthony because your father thought Tony was a bit like Timmy. But then we all found it was a bit *too* much like, so we called you Anthony.'

She stared at him, frowning, and Anthony gazed back, interested but not surprised. Then his granny went on, 'Nobody here knows you had an elder brother, and yet you call your friend by his name. Somebody must have told you about him. Now, who was it, Anthony? Who told you?'

Anthony wondered why she was making such a fuss about something so simple. 'Timmy did,' he said. 'He's my brother.'

the visitor

Ian Serraillier

A crumbling churchyard, the sea and the moon;
The waves had gouged out grave and bone;
A man was walking, late and alone . . .

He saw a skeleton on the ground;
A ring on a bony hand he found.

He ran home to his wife and gave her the ring.
'Oh, where did you get it?' He said not a thing.

'It's the loveliest ring in the world,' she said,
As it glowed on her finger. They skipped off to bed.

At midnight they woke. In the dark outside,
'Give me my ring!' a chill voice cried.

'What was that, William? What did it say?'
'Don't worry, my dear. It'll soon go away.'

'I'm coming!' A skeleton opened the door.
'Give me my ring!' It was crossing the floor.

'What was that, William? What did it say?'
'Don't worry, my dear. It'll soon go away.'

'I'm touching you now! I'm climbing the bed.'
The wife pulled the sheet right over her head.

It was torn from her grasp and tossed in the air:
'I'll drag you out of bed by the hair!'

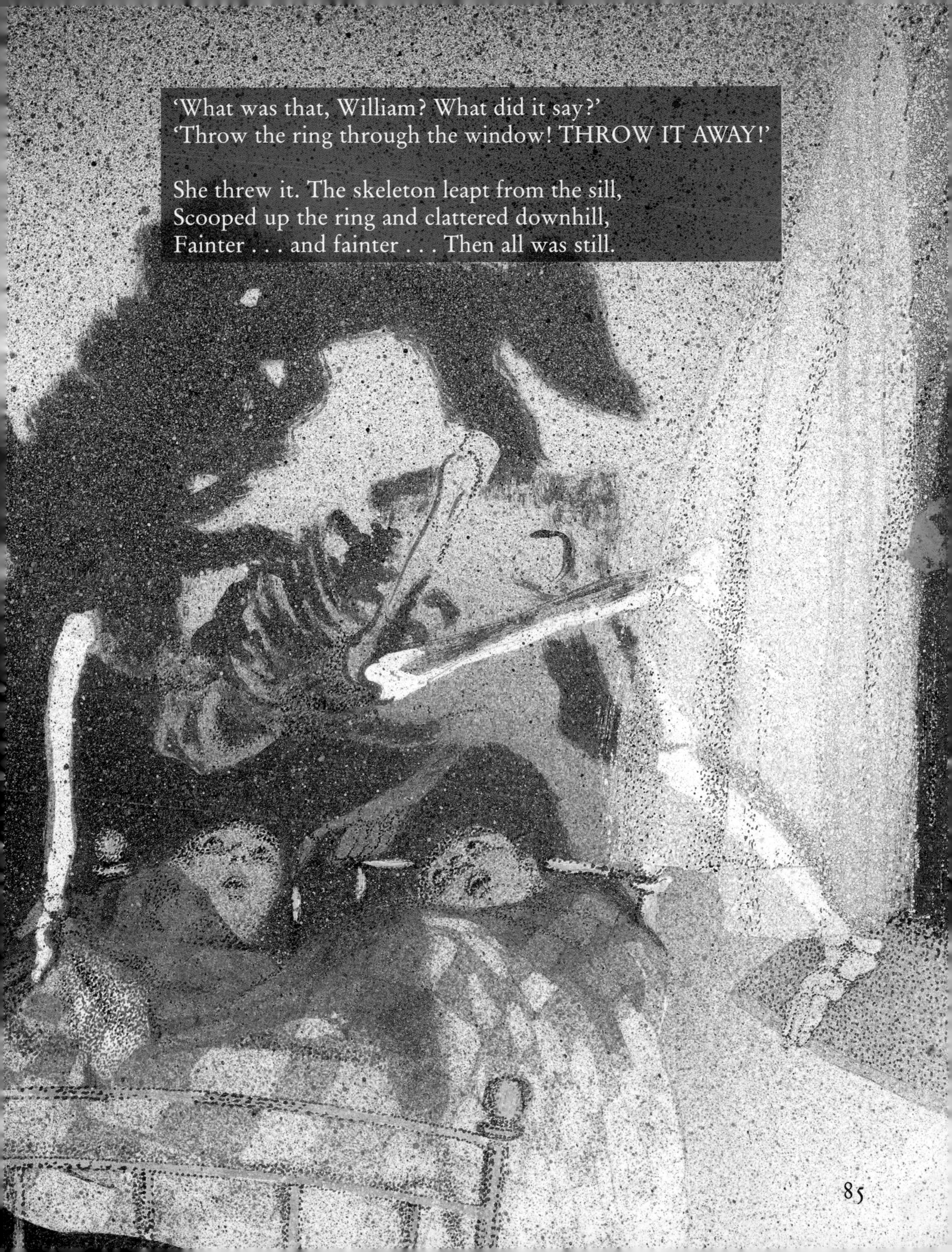

'What was that, William? What did it say?'
'Throw the ring through the window! THROW IT AWAY!'

She threw it. The skeleton leapt from the sill,
Scooped up the ring and clattered downhill,
Fainter . . . and fainter . . . Then all was still.

The Big Black Umbrella

Maria Leach

The big black umbrella stood in the corner of Mary Simmons's room. When it rained she used it, and it kept her dry. But nobody ever borrowed it. No one would think of asking to borrow it. If anyone had, Mary would have refused. It was not for lending.

Mary Simmons was a Black cook and washerwoman in Charleston, South Carolina, and she lived in a small house in a forgotten old neighbourhood bordering a forgotten old burying ground for the poor. On Sunday afternoons Mary liked to sit by her window and take a cat nap.

One drizzly Sunday afternoon she was wakened from one of these cat naps by the sound of singing. What was the song? It seemed to be coming from the nearby burying ground. What were they singing?

> No more rain gwine to wet you—no more
> Oh, Lord, I want to go home—go home—

This was the old, old chant for the dead on the way to burial.

Mary Simmons loved to sing, and she never begrudged her wonderful voice to the dead. She left her room and walked the short distance to the old burying ground where she could see a group of people gathered together for the burial service. She joined the mourners, singing with all her heart the familiar comfort chant for the dead:

> No more rain gwine to wet you—no more
> No more cold gwine to cold you—no more
> Oh, Lord, I want to go home—

Soon it began to rain hard. The wind was cold, but the singing never faltered. Mary took off her hat and was about to slip it under her skirt, when a tall man in a black coat stepped beside her and

said, 'Sister, a song like yours needs a dry cover.' And he put his umbrella into her hands and stepped away.

So Mary sang on in the rain, out of her full heart, the old chant which South Carolina coastal Blacks used to sing for their dead.

When the singing was over, Mary could hear the deep voice of the preacher somewhere in the crowd of people pronouncing the benediction. She bowed her head. '. . . and the blessing of God Almighty abide with you all forever. Amen.'

The Amen was like the shout of an army of voices. The ground shook. But when Mary raised her head and looked about, she saw that she was standing alone in the rain in a little graveyard that had been forsaken long ago. There had been no funeral in that grass-grown place for years and years.

She was frightened, and she hurried, stumbling, from the unkempt spot. When she got to her own door, she realized that she still had the umbrella in her hand. She looked back down the darkening road for the tall man in the black coat, but there was no one in the road. There was no one anywhere in sight. Rain and twilight together created their own grey emptiness.

Puzzled and afraid, Mary wondered what would happen next. She thought perhaps some ghost would come in a clap of thunder and snatch the umbrella out of her hands. But nothing happened. So she walked into her room and stood the umbrella in the corner. There it stood for years. Mary used it in wet weather, and it kept her dry. But it was not for lending.

They were sitting round trying to scare one another.

Jim told them about the creature in the Fens which his dad had seen moving about at night, but the others just laughed and said it was only Jim's dad trying to make him keep away.

Cathy told them about the woman who found a hairy toe and took it home to stew but the monster came after her and got it back. Jim and Kev had heard the story before and said that it was just a story.

Jeanie told them about the miller's boy who had been attacked one night by a huge black cat and had managed to cut off one of its paws before it vanished. The paw turned into a hand and the next day the miller found that his wife had one hand missing. She was burnt as a witch. Jeanie said that if you went to the old mill at midnight and said the right spell the witch would appear.

Kev said he didn't think there was any point in making a witch reappear, but if you went in the graveyard late at night and lay down on a grave the ghost of the dead person would grab you and drag you under.

'That's silly!' said Jeanie. 'What would it want to do that for?'

'I don't know, but it does, anyway.'

'Well, I don't believe it.'

'Huh! I bet *you* daren't do it.'

'I'm not afraid of anyone who's dead,' said Jeanie. 'They're just dead, that's all.'

Jim interrupted. 'It's not just anyone. It has to be a fresh grave. Someone who's just been buried.'

'You mean like old man Crosby?' Cathy asked. 'He was buried today.'

'Yeah,' said Kev. 'Dare you to go and lie down on old man Crosby's grave tonight.'

'Old man Crosby,' Jim said softly. 'He always was rather sweet on you, wasn't he?'

'He was a disgusting old man and I'm glad he's dead!'

'Yeah, but I bet you daren't go and lie down on his grave tonight,' said Jim.

'Dare you,' said Kev.

Jeanie was quiet.

'Go on,' Cathy said. 'You show them. You don't believe in ghosts anyway.'

'Well?' Jim looked at her, knowing his sister would back out.

'All right, I will then.'

'How will we know if she does it?' Kev asked.

'Well, you could all come with me, then you'd know.'

'Oh no we won't!' said Kev. 'You have to be alone. It doesn't work if you're not alone.'

'Sounds to me as if you're the one who's scared,' Cathy said.

Kev grinned at her. 'You bet!' he said. 'You wouldn't catch me anywhere near there tonight. Not with old man Crosby waiting to drag you down and spend the rest of eternity with him.'

'That's just plain stupid!' Jeanie shouted. 'Of course I'll go. We're not all as scared as you.'

Kevin winked at Jim behind Jeanie's back.

'I'll make sure she goes out after everyone's asleep,' Jim said.

'I don't need your help, thank you.'

'And you can borrow my knife,' said Kev, handing her his sheath knife. 'You can stick it in old man Crosby's grave tonight, then we'll know you got further than the gate.'

'You're not supposed to have that,' said Cathy.

'Oh, give over! She's not going to do it anyway.'

But Jeanie put the knife in her pocket and walked away.

Standing outside the churchyard gate at half past eleven that night, Jeanie wasn't feeling very confident. She shouldn't have let Kevin and her brother trick her into accepting a dare. She didn't believe in ghosts and she didn't believe in Kev's story about being pulled into a grave, but in that case what was she doing here? She didn't have

to prove anything. She would have gone back but she knew that if she did they would keep on at her about being scared. She *wasn't* scared, but now she had to go through with the whole silly business.

And I bet the damn gate squeaks, she said to herself.

It did.

Jeanie shivered and pulled her coat tight round her. She could just make out the gravel path snaking between the shrubs and abandoned tombstones and followed it to where she knew old man Crosby's grave would be. She half expected Kev and her brother to jump out at her dressed in sheets, and she determined not to be startled. *Oh, hello*, she would say, *decided to come up for some fresh air, have you?* But they didn't appear.

Old man Crosby had been a vile old man: bad tempered, dirty, often drunk, he would sit on the bench outside the school watching the girls in the playground and shouting comments at them. From time to time he was moved on by the police or, less often, the headmistress, but he was soon back. He had been found dead in a shed on the allotments at the weekend and been hurriedly buried, unloved and unmourned, less than twelve hours before.

She found his grave, marked by a small bunch of ragged flowers, and peered round for any sign of the boys. There was none. Determined now to get everything over and done with as soon as possible, she took Kevin's sheath knife from her coat pocket and knelt down on old man Crosby's grave. She plunged it hard into the soil and started to move away.

She was caught. She tried to stand, but was pulled down again. She tried to roll to one side but was held. Terror gripped her. She moaned as she struggled again to escape, but couldn't. She remembered the time old man Crosby had caught hold of her when she crossed the allotments late one evening, and screamed. Nobody heard her.

Jim was at Kev's house before breakfast.

'She's not back,' he said, his voice shaking.

Kevin started to make a joke, but looked at Jim's face and thought better of it.

'What do your parents say?'

'They don't know yet. They think she's still in her room.'

'Are you sure she isn't?'

Jim looked at him. 'There's only one place she can be. Are you coming or,' he looked hard at Kevin, 'are you scared that old man Crosby'll get you?'

They found Jeanie's body sprawled across old man Crosby's grave. She was held there by Kevin's sheath knife, which she had driven deep into the ground through the bottom of her coat.

My Great-Grandfather's Grave-Digging

Susan Price

This is a true story because it happened to my great-grandfather.

His name was Jody Price, and he was a coal miner until his cough got so bad that he was sacked. He had chronic bronchitis and used to cough himself unconscious, and you can't cut coal when you're unconscious. So he was no use and was turned off.

All that coughing had given him a deep, wheezing, groaning voice that sounded as if it was coming from a mile deep, from the coal at the bottom of the pits. You could always tell when Jody was coming because he soon got out of breath, and then you could hear him panting and wheezing yards away.

After he left the pit he got a job as a gravedigger. He would dig a grave the day before it was needed, get his pay from the verger, and then wander into the Wander Inn. He'd have a couple of halves in there, and then he'd drop in at the Dew Drop Inn and have another couple. He'd play a game of dominoes, have a chat, and then go wheezing home.

He always took a short cut home through the graveyard, and one dark night he fell into the grave he'd dug that afternoon. He lay on his back at the bottom, looking at his feet waving over his head and wondering how they'd got there, until he realized what had happened.

It wasn't easy getting out because the grave was six feet deep, and Jody was only five feet tall—and he was drunk, and the sides of the grave were soft and muddy. He kept slithering back to the bottom, plastered in mud. If he'd had his shovel he could have stood on that, but it was up on top, stuck in the pile of dirt that he'd dug out to make the grave. He got tired and breathless, and made up his mind to sit in the grave all night, until somebody came along to give him a hand out in the morning.

So he hunched himself in a corner of the grave, hugged his hands under his armpits, and tried to forget about the mud and the cold wet. He dozed, but nearly had the life frightened out of him by a sudden rush, thump and a yell. He peered about in the dark. There was a lot of cussing and muttering going on close by, and something was blundering about, thumping into the sides of the grave. Somebody else had come tipsy through the graveyard and had tumbled in.

Jody didn't say anything to let this other man know he was there. How was Jody to know who it was, in the dark? The kind of people who fall into graves at that time of night aren't always the kind of people you want to know.

The stranger kept leaping at the sides of the grave, and falling back with a thud and lumps of mud. Soon he was gasping for breath, but he was no nearer getting out. Jody always did a good job and any grave of his was a good six feet deep, and a few inches over. Jody knew the stranger was never going to get out, if he tried all night. So, in the end, he felt sorry for the man, and he spoke up from his dark corner, in his deep, groaning, wheezing voice. He said:

'You won't get out of my grave. I been trying for ages, and I can't.'

But before Jody had finished speaking, that other man was out of the grave and halfway across the churchyard.

'He could jump like that,' Great-Grandad said, 'and yet he never turned back to give me a hand. He never even said goodnight. Ain't there some camels in this world?'

Poem

Hugh Sykes Davies

In the stump of the old tree, where the heart has rotted out,/there is a hole the length of a man's arm, and a dank pool at the/bottom of it where the rain gathers, and the old leaves turn into/lacy skeletons. But do not put your hand down to see, because

in the stumps of old trees, where the hearts have rotted out,/there are holes the length of a man's arm, and dank pools at the/bottom where the rain gathers and old leaves turn to lace, and the/beak of a dead bird gapes like a trap. But do not put your/hand down to see, because

in the stumps of old trees with rotten hearts, where the rain/gathers and the laced leaves and the dead bird like a trap, there/are holes the length of a man's arm, and in every crevice of the/rotten wood grow weasel's eyes like molluscs, their lids open/and shut with the tide. But do not put your hand down to see, because

in the stumps of old trees where the rain gathers and the/trapped leaves and the beak, and the laced weasel's eyes, there are/holes the length of a man's arm, and at the bottom a sodden bible/written in the language of rooks. But do not put your hand down/to see, because

in the stumps of old trees where the hearts have rotted out/there are holes the length of a man's arm where the weasels are/trapped and the letters of the rook language are laced on the/sodden leaves, and at the bottom there is a man's arm. But do/not put your hand down to see, because

in the stumps of old trees where the hearts have rotted out/ there are deep holes and dank pools where the rain gathers, and/if you ever put your hand down to see, you can wipe it in the/sharp grass till it bleeds, but you'll never want to eat with/it again.

Bronwen and the Crows

Alison Leonard

First there was one crow, and he was beautiful. Then came the other crows.

Bronwen, who was ten, had been looking for a friend. Now she'd got one—her own pet crow.

Imagine hoping that a little sister would be a friend! Little sisters were useless. They whined, and played with dolls, and ran inside to Mum. But a crow could fly, and bring you twigs from faraway fields to build dens with, and you could feed him with slugs.

Bronwen had found him in the spring, a tiny featherless black scrap, squeaking pathetically at the foot of the oak tree behind the big barn. His beak was almost as long as his body, and sharp as the corner of Daddy's wood-cutting axe. She picked him up tenderly, took him inside and laid him on a soft cloth in a shoe box. She offered him a morsel of raw mince, and it vanished down his hungry throat.

Mum said they must keep him well away from little Elaine. Only last month, a crow had nearly pecked a toddler's eye out, down in the valley beyond Llangollen.

As Bronwen's crow grew, his feathers became glossy, blue-black, the colour of the night sky. He learnt to fly, but he never flew far from Bronwen.

She christened him 'Nos'. It meant 'Night' in Welsh, and it rhymed with 'close'. 'Night' would rhyme with 'fight', Mum said. She was always warning Bronwen not to fight with children at school.

In the mornings, while Mum was trying to get breakfast into Elaine, Bronwen would run out to Nos and find him hopping round Tudor the donkey, or perched on the gatepost. She could tell him all the things she'd like to do to three-year-old sisters, and he'd nod: 'Yes, yes. If I had a sister I'd do that too.'

Mum discovered Nos pecking in the lettuces one day, and clapped her hands to scare him away. But Bronwen knew what he wanted: not the lettuces, but the slugs. She came back when Mum had gone indoors and eased apart the leaves to pick them out for him. Then, borrowing Elaine's toy knife and fork (which used to be *her* knife and fork), she cut the big fat slugs into bite-sized pieces and laid them out on the path for him to gobble up.

Even when he was big enough to swallow the slugs whole, Nos still liked to wait patiently beside her while she sawed away with the blunt knife. He ate the pieces one by one, looking up at her between bites with his bright friendly eye.

Mum said that when he was fully grown, they'd have to let him fly away. 'Crows are dangerous things,' she said. 'You never know what they'll do with that great beak of theirs.'

But Bronwen knew Nos would never leave her. Mum said, 'He'll get lonely without other crows.' Rubbish, thought Bronwen. He's got me.

Then the other crows came.

She first saw them round the rabbit hutch. Nos was friendly with the rabbits, especially Black Rabbit. If Black Rabbit was munching a cabbage-leaf, Nos would poke his beak in and grab half of it. Then he'd fly up into the oak tree and set about tearing the leaf into shreds, with a puzzled look as if to say, 'Can this be *food*?'

It made Bronwen angry to see him by the rabbit-run, with Black Rabbit nuzzling his beak through the wire. Nos was *her* friend, no one else's. And Black Rabbit belonged to Elaine. He was born on Elaine's first birthday. She'd toddle out on her little fat legs with rabbit food clutched in her sticky hands, calling, 'Back Yabbit! Back Yabbit!'

Ten crows, huge and black, flew down that morning. Bronwen watched them from her bedroom window. Their wings moved like a slow-motion ballet, the tips turned upwards in a question-mark. They settled on the grass, arranging themselves in a circle round the rabbit-run. Then they stood, very still.

Something happened to Bronwen's breathing. Panic filled her throat, her chest felt tight.

At breakfast she didn't say a word, but Mum didn't notice. Dad was busy with a sick cow, and Elaine tipped her Weetabix on to the floor. After breakfast, Bronwen went to the sink and got some limp radish-tops and carrot-peelings from the rubbish bowl. She called to Mum, 'I'm just going to feed the rabbits!'

The crows stood in their circle. They hadn't moved.

She'd never seen birds so still before. Rabbits could act like statues if a hawk was hovering above, and sheep looked dead when they were asleep. But birds . . . perched on the grass . . . waiting . . . Her chest felt tight again; it hurt. She threw the rabbit food down on the grass and ran, panting, back to the farmhouse. Then, from the safety of the doorway, she looked back.

The crows had taken off on vast beating wings and, still in circle formation, were rising into the air like a witches' ring. Nos, her friend, was nowhere to be seen.

That afternoon Nos was back, perching on his gatepost as usual. He came flapping over to Bronwen as she came out of the house, and led her to the bank at the side of the lane where he'd got one of his hidey-holes for slugs. Now he was nearly grown up, he found most of his own food, but he still liked to be fed with slugs. He would fill his beak and fly off to bury them in special hiding places, to save for when he wanted a tasty snack.

By bedtime Bronwen had forgotten about the black circle of silent crows. But she dreamt about sharp beaks stabbing like daggers at a baby's cot. She woke up wide-eyed with excitement and fear.

Next day Mum wanted Bronwen to help with the baking. Elaine was playing out in the yard, near to the shed that they called the Play House. Mum kept looking out of the window or the door to make sure she was all right.

In the middle of making some Welsh cakes, Mum needed some more flour. 'Keep an eye on Lainey, Bron,' she said, and went off to the store-room where the big flour-bags were kept.

'Keep an eye on Lainey indeed,' thought Bronwen. She was scraping the mixing-bowl clean of chocolate cake mixture, letting the creamy taste roll round her mouth and slide down her throat. 'The crows can keep an eye on Lainey.'

A shrill scream from the yard pierced her chocolatey daze.

Elaine! Bronwen was out of the door like a terrified cat. It might have happened—the crows might have got Elaine! In that split second, she thought, 'Good riddance to bad rubbish!' and 'No, no—they mustn't hurt Lainey!'

Elaine had dragged her doll's cot out of the Play House. She was holding the doll in both hands and waving it up and down, screeching as if lions and tigers were roaring around her. She wasn't even hurt.

Bronwen was furious. 'Shut up, Lainey! What's the matter?'

'Dolly! Arms all gone!' shrieked Elaine. She thrust the doll towards Bronwen.

Bronwen looked. One of the doll's arms had been wrenched off, and where the arm used to be, there was a gaping hole.

Mum came rushing out. 'Bronwen! What on earth—? Darling—what is it?' She knelt down to cuddle Elaine.

'Dolly's arm's gone! Very very gone!'

'But the dolly was fine when you put it away yesterday!'

Elaine struggled free. 'There, there!' She pointed at the grassy bank at the edge of the yard. Poking out of a muddy hole, where Nos had one of his foodstores, was a pink doll's arm.

Mum looked straight at Bronwen.

Tears of fury rolled down Bronwen's face. 'It wasn't me! It wasn't, it wasn't!'

'Well, who was it, then?'

'It—it must have been Nos!' As soon as she said it, she knew what Mum's answer would be.

'I don't see how it could have been Nos. But if it was . . .' to punish Nos was to punish Bronwen '. . . that bird will have to go.' Mum swept Elaine into her arms and, murmuring soothing words, carried her into the house.

Bronwen looked up. There, high up on the bank beyond the yard, perched a row of ten, black, silent crows.

She gasped, ran in through the door and up to her room. For the rest of the morning she lay on her bed, crying and raging against Elaine. At dinner-time she wouldn't speak. In the afternoon she went in search of Nos, but couldn't find him anywhere.

The next day was Sunday. When they got in the car to go to church, Nos flew down. He flapped round them in his friendly way, and Mum pursed her lips. Bronwen thought, 'If Mum sends Nos away, I'll never forgive Elaine.'

In the afternoon, Elaine was playing round Dad's feet while he read the paper, and Mum said, 'Bron, you can help me cut Tudor's toenails.'

Trimming the donkey's horny hooves was a job Bronwen always enjoyed. Today she was suspicious. Was Mum softening her up before delivering the final blow about Nos?

Nos flapped down to watch them. He was alone. Why did Nos disappear when the scary crows came?

Mum got the big clippers, and Tudor the donkey placidly lifted first one hoof, then the next. Bronwen's job was to collect up the trimmings that shot all over the grass.

She heard the crows before she saw them. Their wings beat the

air in a steady, threatening rhythm. At the sound, Nos took off. He flew up to the other crows, and Bronwen, staring, couldn't tell which of them was her own special friend.

Mum was busy with Tudor. One of his back hooves had a sore patch, and she had to be careful.

Bronwen looked over to the house, and saw Elaine. She'd come out of the back door and was toddling down the path. Dad must have fallen asleep, the Sunday paper over his face.

The crows settled beside the path. They stood, silent as before, in two half-circles, the path between. Elaine was coming towards them. She trusted them. They were like Nos, they were only crows.

Bronwen stared. She dropped the hoof-clippings from her hand and stood fixed to the spot. Elaine was singing 'Dum! Dum! Dum!' each time her welly-booted feet hit the ground.

The crows opened their beaks at her approach.

'Dum! Dum!' Her little boots were within one pace of the crows.

Bronwen could see it—the savage beaks ripping Elaine's little denim dungarees—her screams—the blood. . .

'Lainey!' Bronwen's terror almost lifted her feet off the grass. Within a second she was holding the stout little body in her arms.

The noise of wings was deafening. As the huge black shapes lifted into the air, Bronwen forced herself to look up at the crows.

One of them separated himself from the circle. It was Nos. He flew round Bronwen as she crouched, clutching Elaine. His bright eyes spoke silently to her: *I'm going, and I'm taking the other crows with me.*

Then he flew up, rejoined the circle of black shapes, and they flapped away, out over the valley.

'Bron! Lainey!' Mum was running towards them. 'What's happened? Are you all right?'

'Yes,' said Bronwen. 'Nos has gone. I told him he'd got to go, so his friends couldn't hurt Elaine.'

'Good girl,' said Mum.

'Dum-ti-dum, dum-ti-dum,' sang Elaine, and she skipped back towards the house. Mum took Bronwen's hand, and they followed after.

Supermarket

Dennis Hamley

Gary, who is nine and small for his age, wakes up from a bad dream about drowning in a wreck-strewn and serpent-ridden sea and finds he feels much better. All through that week he has been away from school with flu: on Monday and Tuesday he was really terrible; on Wednesday he was merely awful; on Thursday he was just weak. But now, Friday, he leaves his choking dream to sit suddenly up in bed and say aloud, 'I feel all right.'

It's silly, he thinks, to go back to school just for Friday. He lies down again and feels even better.

Mum agrees before she goes to work. As she makes sure he will be safe and warm on his own, she says, 'You can get some fresh air over the weekend. And for a start you can come late night shopping with me this evening at Safebury's.'

Mum comes home at one o'clock and gives him some soup. She sees he is pale but lively and goes back happy. But when she has gone, Gary feels strangely tired and goes back to sleep, this time to a dream of mazes and being lost for ever though the way out is just round the corner.

When he wakes again his mind spits and crackles with memories of his dreams and the strangeness of daytime sleep.

Mum returns: they eat: they pile into the ancient Renault 5 and chug to Safebury's. It is dark now: the streetlamps are orange and Safebury's car park is vast but full. Parking the Renault takes longer than their journey. They get out to the clash of trolleys hurled against each other by leaving shoppers and dodge the long steel snakes of reclaimed trolleys, their guides at the rear slumped like galley slaves.

Safebury's revolving doors move slowly. When Gary pushes to hurry them up, they stop. He feels guilty. Inside, Gary and his mother walk through a wall of heat and light which makes Gary blink. Once through the barrier, Mum becomes single minded. Her trolley fills, first with fruit, then with vegetables. The rasp of plastic bags she tears off the rolls becomes a rhythm.

The aisle is crowded. Trolleys lie in all directions like the ships of a scuttled fleet. But his mother negotiates them and the people clinging to them like a confident pilot and Gary is glad to follow.

They push their way to the delicatessen counter. Mum looks at the rows of trays of cooked meats, samosas, quiche, and pies and says, 'Get me a ticket, Gary.'

Gary looks to right and left. He sees that a sinister little metal monster with a shiny silver neck rears near by out of the floor like a sea serpent. Its pointed head has a sharp, clicking mouth in which people are putting their hands. Gary, remembering his dream, brings himself to do the same and pulls on the monster's thin little tongue. The ticket he now holds has a number written on it—21. There is a shout of 'Next, number four,' and, high above him, is a buzz and 04 appears in white on a black panel.

Gary gives his mother the ticket.

'I've got to wait hours,' she says. 'Gary, be a dear and pick up a few things for me while I choose what I want here.'

'I like the turkey pie with cranberries on top,' says Gary.

'We'll see,' says Mum. 'I want you to get me one strawberry jam, one blackberry and apple, one plum and two jars of orange shred marmalade. Fine cut. Can you remember that?'

'Where are they?' asks Gary.

'Somewhere. It tells you on the notices over the shelves.'

Mum turns to the counter and forgets him.

Gary looks up. Yes, there are signs hanging from the ceiling to show where things are.

PASTA	CRISPS/NUTS	CANNED MEAT	TEA/COFFEE	WINES/SPIRITS
BISCUITS	CONFECTIONERY	CANNED FISH	PICKLES	PRESERVES

Where is the sign saying JAM AND MARMALADE? He turns to his mother again; she is swallowed up in the queue. He must search the shop without help. He feels cast off into a strange sea.

Gary walks down the first aisle. Or he tries. Soon he is *fighting* his way down the aisle. He struggles past immoveable, shouting people whose hands are on a level with his nose. His knees and stomach are buffeted by the ends of trolleys. Handles hit him on the chin. He cannot see faces: hands snatching avalanches of goods wave across his eyes. Soon he has no idea where he is. People obscure the shelves. Only by looking at the neon-lit ceiling can he track his path. It is his maze dream.

A tinny voice sounds over the racket.

'Good evening, ladies and gentlemen, and welcome to Safebury's.'

Gary pauses. He waits for the voice to say, 'Gary, the shelf you want is half-way down the next aisle.'

Instead, it says, 'Haddock and cod are reduced on our fresh fish counter.'

Gary feels deserted. He can only push on.

The movement of people stops; no struggling; no pushing. What can have happened? Gary stops too. It takes him some time to realize he is in a checkout queue which stretches right down the aisle. He escapes from it in time and finds himself at the point where the shelves end, with nothing for it but to turn and fight his way up the next aisle. He also realizes the only way he will ever tell what is on the shelves is by seeing what people put in their trolleys. If he goes right up to the goods he will have twice the journey and probably be flattened to death on the very articles he is looking for.

The next aisle is worse. Before, nuts and crisps were landing softly in the trolleys. Now, tins are flying in and crashing against thick steel wire so as to make Gary's head spin. Tinned fish, tinned meat, tinned beans, tinned peas: waterfalls of metal. They are hurled to the bottom like wrecked hulks spinning towards the depths of the sea.

Gary is dizzy. He is in his drowning dream. He pushes on.

Next aisle. Bottles and jars. Close, surely. But they are of coffee and pickles, Marmite and Bovril. The noise is different. Teabags provide a respite. But now Gary is lost. The drowning dream and the maze dream have combined and panic rises in him. He wants—he needs—to scream.

Every shelf surely has been passed. He sees whole walls of whisky and cans of beer. He sees rank on rank of lemonade and cola bottles. He sees no jam, no marmalade. He is lost in a jostling, suffocating, trackless wilderness. He wants to sink to the ground and cry. But if he does he will be trampled underfoot; he will be drowned; he will be left for dead on a path with no beginning and no end.

And then, without warning, the crowds thin. The shelves in front of him appear like stage scenery when the curtains open. Rows of jam, marmalade, honey. Miracle. But first he looks up. What word hangs overhead that he should have looked for? PRESERVES? He shakes his head, mystified.

Then his heart sinks. He has no basket or trolley. How can he carry five jars?

There is only one way. He selects a jar of strawberry, a jar of plum, a jar of blackberry and apple and two jars of orange shred marmalade, fine cut. He clasps them to him in a row against his jersey, his arms folded across his chest. Cautiously he turns round. Yes, he is free of people; they surround him some way off on three sides; it is as if he stands in a clearing in a forest. He steps confidently across it.

Without warning, powerless and out of control, he is falling. He lands face down on the floor. The jars break. Broken glass cuts his hand. Strawberry, blackberry and apple and plum jams join and spread gorily across the floor. The marmalade ends up glassless as two solid little orange cylinders.

The noise dies away. Everyone looks at Gary. A woman in a smock, carrying a huge wet mop, bustles up.

'I told everybody to stay out of the way till I cleared the mess up,' she screams.

Gary scrambles to his feet. Too late he sees why there was a gap in the crowd. Someone has spilt a whole bottle of vegetable oil. He has slipped in the slimy puddle. The jam and the oil together now make a mess beyond a wet mop to clear.

Gary does not wait to find out whether people are angry or sorry about him. He clasps more jars to the now filthy and bloodstained front of his jersey and melts back into the crowd.

Again the suffocation. Again the hands at the level of his nose: fat hands, thin hands, smooth hands, hairy hands, knobbly hands. Once, alarmingly, a hook. He looks up to see the faces: thin, fat, made-up, stubbly, bearded, shiny. Mouths open and shouts of irritation pour out of them. Nobody smiles, nobody laughs.

Gary has to cling tight to his jars. He moves even slower than before. Soon, he stops altogether. He is wedged in a tide of trolleys, shoppers and shelf-fillers and can only go where the tide goes; slowly up one aisle, slowly down the next. He almost wants to sink to his knees and sleep while the trolleys run over him. He will never see his mother again: he will perish on this mad battleground. Both his dreams have come true at once only this time there is no comforting sleep to wake up from; he is *in* them and there is no escape.

Two trolleys wedged at an angle nudge him backwards into a corner. His shoulder is forced against something hard. Without moving his arms, he turns.

Joy. It is a landmark. The metal monster with the paper tongue. He has returned. His mother will be here, waiting.

He looks for her. She is not here. He hears a buzzing noise overhead. He looks up at the number which appears. 86.

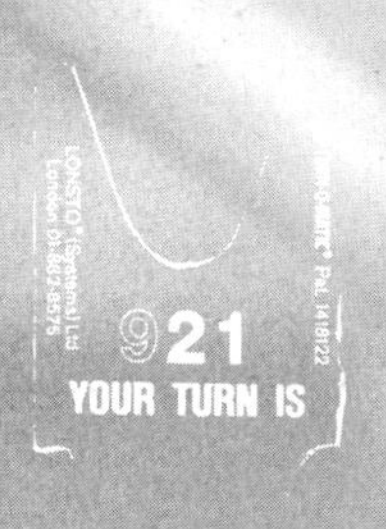

Now he really does sink to the floor in despair. She will have gone, long ago. She will have disappeared from his life. He is left, hopeless and sticky, on his own in a madhouse. The dreams have taken over completely.

'And the next please.' A man's voice. Gary straightens up and looks across the counter. A young man in a white coat with a white boater stuck on his head slices cheese.

Cheese? Who wants cheese? Not his mother; she hates the stuff. It was ham, quiche, samosa, and with a bit of luck some turkey pie with cranberries on the top that she was after. So, in this appalling place they even mix up all the things they sell just to confuse him. Despair strikes again. He turns savagely on the misleading metal monster, wanting to uproot it and smash its head in.

But he doesn't. Though his arms, stuck to the jars, ache, he is scared of moving them. He stands hopelessly by the monster. He cannot do anything; he cannot think anything.

But wait! A thought has entered his mind. Perhaps they haven't shifted all the goods round. *Perhaps there are two metal monsters.*

Cautiously, like someone trapped on a high narrow ledge inching his way to safety, he shuffles in a straight line along the front of the counter. He ignores the disgusted cries of, 'Ugh! Horrible, dirty little boy.' Soon, the articles under the glass stop being cheese.

And (joy again) here he stumbles into another little metal monster—and (oh huge relief) under the glass he sees a big coiled sausage. And overhead he hears another buzz and sees another number—and it is 21. And he hears a voice say, 'Next please,' and in answer is another voice saying, 'Eight ounces of ham, two slices of spinach quiche, and four portions of that turkey pie with cranberries on top.'

And it is his mother. And suddenly the dreams roll away. And she sees him and says, 'Well done, Gary. You timed it nicely. Aren't you in a mess! What happened? Never mind, you forgot to take a basket. But you didn't forget what I wanted.'

'No,' says Gary with pride. 'I remembered. And I could carry these few jars all right.'

Already the terror is fading from his mind.

THE CAVE

Gregory Harrison

I told the boatman to go deep,
Deep in the hollow cave
Where the black menace of
Each lifting wave
Licked with wet tongue and froth of lace
This barnacled and shadowed place.

I used to come here as a boy
And here at low tide felt the race
Of pulse
And throbbing in my throat,
Standing upon a shelf
Above the dark and sinuous moat
Of water,
Knowing that if my feet were caught
By mischance in a cleft of rock
I should be struggling in a lock
Of rising water.
The very thought
Of choking in that monstrous swell
Made me wheel round and yell
With panic.
I used to clamber, slide and slither back
Along that grey, forbidden track
And sob in safety on the tufts of thrift
Watching the cheated rollers shift
Slyly beneath my feet.

This day, in someone's boat
I catch again the terror of the place;
I clutch the collar of my coat,
And tell the boatman curtly, 'That's enough.'
The sweat is cold upon my face.

The Hawk

John Gordon

The coffin tilted as it slid below ground, and the woman could not bear to look. She raised her eyes and, in the sharpness of the bright sunlight, saw the hawk at the edge of the churchyard.

High on a dead branch, it spread its shoulders as it dipped its head to the vole at its feet. The hawk's bones and tendons were a machine under the feathers, and when it nodded its head a hook pierced the vole's skin and tore it. The bird shuddered, tearing backwards, and when it lifted its head, a few red strings of meat hung from its beak.

The mourners had bowed their heads as the coffin nudged the earth, but the woman looked above them. There was the width of the churchyard between her and the hawk but it caught the movement of her head and held itself still, not feeding. She hated it. It crouched over its kill like a cat, and its eyes were the same as a cat's that had just killed, not fierce, but round and peaceful, content with what it had just done. And her son slid underground.

The sob that escaped her was as harsh as a cough, and the bowed heads swayed, embarrassed by it, wanting to escape from her grief. Her husband's grip tightened on her black gloves, and she squeezed in return, wanting to tell him that it was not only the anguish that made her cry out. She wanted to send the hawk flying.

Her cry had made it crouch lower, cat-like, but then it tossed its head, gulping meat, and in the same movement launched itself from the branch and in a lazy upward curve, trailing the dangling paws of the vole, vanished behind the church tower.

The woman had succeeded in driving it away but, with the disappearance of the bird, her grief grew unbearable and her weight bore down on her husband's arm as the first earth trickled on to the coffin.

The mourners had gone when the two girls came to the churchyard gate. The gravediggers had taken away their tarpaulins and planks and ropes, and had shaped the mound and placed the wreaths to hide the earth so that Paul Simpson's grave was no more than a mound of bright flowers. The girls caught only glimpses of its brightness as they threaded their way between the grey headstones of ancient, settled graves.

'It look like a bedspread, don't it?' said the younger girl, whispering. 'Poor little Paul.' She was ten, and Paul had been the same age, but he seemed very small now that he was alone among all these long graves.

'He have some lovely floral tributes,' said the older girl, and they held hands because they were both afraid.

Sarah, the younger girl, freed her hand and pointed. 'Who do you reckon sent that big one? Must be his Mum and Dad.'

'That's from the school, dippy. That's from us.'

'Oh.' Sarah's mouth went very small. 'I forgot.' Her mouth remained almost invisible, but her eyes were large under her wispy fringe. She clutched the bigger girl's hand again, tighter. 'I thought his Mum's would be biggest.'

'It don't have to be.' Bridget's dark hair and dark eyes dominated the smaller, pale girl. 'It don't have to be no more than just one flower. In fact just one flower might be best, with no leaves, nothin'.' The idea appealed to her and brought back the phrase she had used earlier. 'That would be a lovely floral tribute, I reckon, just the one rose all by itself.'

'Is that it, then, that little tiny bunch without no cellophane?' asked Sarah.

They crouched and tried to read the card without touching it.

'That's his cousin, I reckon,' said Bridget. 'That's only a posy, so it wouldn't be from somebody grown up.' The card was at an

awkward angle and she was reaching to turn it when Sarah suddenly stood up, pulling her back.

'What did you do that for?' Bridget glared at her. 'There ain't nothing wrong with seeing who sent 'em.'

'It ain't that,' said Sarah.

'What is it, then?'

Sarah hung her head, not wanting to speak, but Bridget jerked her arm so that she looked up. 'It's all them flowers,' said Sarah. 'I don't like the smell.'

'That's a lovely smell,' said Bridget. 'Ever so nice.' She breathed it in.

'Paul hated it.' Sarah had lowered her head again so that her fringe hid her eyes. 'He always did say flowers smelt like funerals. I remember.'

'Paul's daft.' But suddenly Bridget realized where she was, and corrected herself. 'I'm sorry. I'm ever so sorry. Poor little Paul.' She sniffed, squeezing out a tear to trickle down her cheek, but Sarah tossed her head to clear the fringe from her eyes and gazed steadily at Bridget's bowed head.

'It don't matter,' she said. 'It don't matter to Paul no more.'

Bridget was still sobbing when Sarah saw a bird shake itself free

of the church tower and make a long glide over the churchyard. A tree held out a dead branch like a crooked finger, and the hawk settled and watched them. Sarah stared back at it. 'It have a flat head like a snake,' she said.

'Snake?' Bridget gasped and lifted one foot, afraid something was sliding towards her. 'What? Where?'

'Up there.' Sarah pointed.

'Oh!' Bridget let out her breath. 'You had me worried for a minute. I thought somethin' was coming at us.'

'He loved them birds.' Sarah had not taken her eyes from the hawk. 'Especially that sort. Them hoverers.' The kestrel, as though it was responding to her voice, blinked and raised its head. 'Oh, ain't he proud!' she said. 'He's got a speckly chest just like a cloak.'

The kestrel opened its beak and a thin sound, like the mew of a cat, came from it. Then it unhooked its yellow feet from the branch and slid swiftly over their heads and out of sight.

'I don't want to be here no more,' said Bridget. Sarah and the bird had reminded her that Paul was still only a yard or two from them, clutched tight in the dark earth. The flowers were only a disguise.

Sarah saw that the bigger girl was unsure of herself and took charge. 'All right then,' she said, 'let's go,' and it was she who led the way to the small gate furthest from the church.

They came out into a back road, away from the centre of the village, and were walking quickly until Bridget suddenly stopped. 'Where you taking me?' she said.

'I just wanted to see where it happened, that's all.'

'You can't!' Bridget could not believe her. 'That's never right to go where Paul's accident was. Not just after he's been buried!'

'Well, we did go and have a look when we first heard about it—and that was on the same day.' Sarah remembered the tiny patch of blood on the kerb. It had seemed too small to show where Paul had died. 'I want to go anyway,' she said.

'That ain't right.' Bridget stepped in front, blocking the way. 'What do you think people will say?' Sarah's head was bowed. 'Look at me, Sarah!' But the fringe turned away, and Bridget sighed, knowing she would have to give in. 'Oh, there's no point trying to argue with you when you're like this! Come on, then,' and they walked on in silence.

The lane curved behind the church, and headed out into the flat fenland. Among the wide fields one house stood by itself. It should have been lonely, but today the track that led to it had a scatter of cars shining in the sun. They were bright enough for a wedding, but it was Paul's house and they had come for the funeral.

'There's still a lot of people haven't gone home.' Bridget was whispering as her eyes searched the outbuildings where she knew Paul had kept his bike. She thought she could see it in a little lean-to. 'So it ain't true,' she said. 'His father haven't got rid of it.'

Sarah could see only the saddle and back mudguard; the front wheel was hidden in shadow. She knew the wheel was buckled because she had seen the bike lying at the roadside as they put Paul into the ambulance.

'What d'you reckon they've done with his shoes?' she said suddenly.

Bridget's head jerked towards her. 'What do you mean, shoes?' she said.

'I was just wonderin',' Sarah said. She had seen one of the ambulance men hand the shoes to Paul's mother, and she had watched his mother pluck at the laces, undoing them, as she waited to climb into the ambulance after the stretcher. Paul had tied those laces, and never would again.

'You're morbid, you are,' said Bridget, 'thinking of things like that.'

But Sarah suddenly jerked her head away from the house and the glittering cars and cried out, 'There it is again!'

'What?' Bridget was startled.

'That hoverer. Look at him standing in the sky like that.' The kestrel had chosen a place in the empty air ahead of them and hung above the grass of the verge. 'Paul loved them hawks,' she said.

'That ain't a hawk,' said Bridget, suddenly remembering what she had been told at school. 'A kestrel's a falcon.'

It made no difference. 'I don't care what you call it,' said Sarah. 'Paul loved them birds.'

The hawk saw the girls, fanned its tail feathers stiffly against the turbulence made by its wings, and stood still as though it rested on an invisible wall in the sky.

'Listen!' Sarah ordered, and its thin cry came down to them.

'I don't like it.' Bridget clutched at her arm as Sarah began to move towards it. 'It might come at us.'

'I don't care if it do.' Sarah shook herself free and began to run. The hawk ceased to beat the air and slid away on pointed wings. 'Wait for me!' she shouted as it came up into the sky and halted again. 'Look at that!' she shouted back towards Bridget. 'It heard me!'

'Stupid!' Bridget came reluctantly towards her. 'That's only looking for mice.'

But Sarah ran on, and the hawk, in smooth bounds like a cat in the sky, went ahead of her. Bridget dropped back and was not within shouting distance when Sarah rounded the bend and came to the bridge.

It was only at that place that the road climbed above the flat land. It climbed the steep floodbank of the river, crossed the bridge, and dropped away steeply on the other side. Bridget hung back. Just over the bridge was the place where Paul died.

Sarah, running up the slope, was shouting at the bird in the sky.

'If I had my bike I could keep up with you. Wait for me! Wait!'

It dipped above the centre of the bridge and hung there until she had climbed the slope, and then it swung out over the water, keeping level with her. She stood still, panting, and held out a hand, but it rowed itself backwards in the air and watched her.

'All right,' she said to it, 'please yourself. I didn't come up here just to see you, did I?' and she turned to look down the slope at the other side of the river.

The road fell away steeply to the sharp curve at the foot of the bank. Far away, flashing like glass beads rolling in the sun, cars sped on the main road, but here there was no traffic and the sound of the road died before it reached her. This was where Paul had come to ride his bike.

More than anything in the world he had wanted to fly, so he had made his bike into a flying machine. She had seen him fly down the slope and then lean into the curve, his wheels juddering on the bumps, before he hauled himself upright for the long glide out across the fens. He would stand on his pedals and fling his arms out, flying.

But once it had not happened like that. He had been riding alone and nobody had seen it, but two children had found him at the foot of the slope, yards away from the machine with its twisted forks and buckled wheel.

Sarah went down to the place. There had been one tiny patch of blood. The gravel had been scuffed since then, but the mark was still there, no larger than the shadow of a hand on the yellow road.

She turned and looked back up the slope. The hawk had come down. It stood on the stone capping at the end of the bridge, tilted forward on its feathered legs. The shoulders of its wings rose above its head and its hooked beak gaped wide as if at any moment it would speak.

For the first time it frightened her. 'Hey!' she cried, trying to scare it away, but it did not move. She waited, not wanting to face the bird by herself, but Bridget had not come on to the bridge and was out of sight. She must have turned back. Sarah and the hawk were alone.

To give herself courage, she spoke to it. 'Did you see it happen, then?'

She heard the bird's breath hiss.

'Did you see it?'

The hawk danced, its butcher claws barely touching the stone.

'Was you there, hoverer?'

It was bowing and dipping. She stood below it, at the point where Paul had seen his last flash of the sky as he flew down the slope. She opened her mouth to ask the question again, but the words did not reach her lips. Suddenly, without the shadow of a doubt, she knew that Paul was still flying. He had died flying, caught up in the sky with the bird, and he had gone on doing it. The hawk knew. It was telling her so. It danced and screamed because it had filled Paul's last blinding glimpse of the sky.

'Paul saw you,' she whispered. 'You was with him.'

The bird was quite still now, but it stared at her with its hooked beak open and she was afraid of it. It knew something else, and it wanted her to know.

'Who are you?' she said, and it hissed at her.

Then slowly, as though she was warding off danger, she held up an arm. She paused, gathering her courage, then sent her voice up the slope.

'Paul?' Her voice was too small, like the rustle of a mouse. She cleared her throat. 'Paul?'

It lifted from the stone parapet, hung for a moment, then came for her. She waited for it, forcing herself to stand still, holding her arm out. It raced its own shadow down the slope like an arrow aimed for her face, and at the last instant she flung back her head. But the hawk's feathers flared, halting it against the air, and its cold claws clutched her wrist.

'Paul?' she asked again.

The hawk's tawny eyes were on her, and she was certain.

'Paul,' she said softly, and with one finger smoothed the top of his head.

No-Good Claus

Jan Mark

Until she learned to read, Stephanie thought that Santa Claus was Santa Claws, and imagined him to be a large cat with a tail and whiskers. When Granny took her to see him at Lewis's she had been disappointed to find a human face behind the false beard. She could see that the beard was false; why didn't he do the job properly and wear false fur? When she hung up her stocking next to Marnie's, on the bedpost, on Christmas Eve, she knew that it would be the real Santa Claws, with real fur, who came down the chimney to fill them, and not the silly man from Lewis's who could not even purr.

But when Christmas time came round again she discovered that the silly man from Lewis's was nearer to the real thing than her own idea had been. Santa Claus, it seemed, was another name for St Nicholas.

'In Germany, Claus is short for Nicholas,' her teacher explained, and told the class that in Europe, long ago, children had left their shoes out on the night of 5 December, for St Nicholas to fill with toys if they had been good, or a bundle of birch twigs to beat them with if they hadn't. Over the years the gifts left by St Nicholas had become part of Christmas, and so had he; *Sankt Nikolaus*, Santa Claus.

When Stephanie hung her stocking on the bedpost that year she felt that some of the magic had gone out of it all, now she knew that her furry Santa Claws was just some old saint called Nicholas.

By the time Marnie went to school Stephanie knew that there were no men or cats involved anyway. Santa Claus was Mum, creeping into the bedroom at midnight to fill the two dangling stockings with sweets and little toys and, always, an orange in each toe. She could not remember how she had discovered this, she just knew, but it was still fun pretending, for Marnie's sake. When people said, 'Christmas is for the children,' she understood. She did not look forward to the time when Marnie too knew the facts, and wished that Dad were still with them so that Mum would have another baby and there would always be someone to pretend for.

Then they went to stay with Granny at Christmas. Stephanie guessed that there was going to be a problem as soon as she went upstairs with Marnie to unpack. Granny had put them in the front bedroom that they always shared when they stayed there, but this was the first time they had ever stayed there at Christmas. Marnie took one look at the two beds and wailed, 'Where shall we hang up our stockings?'

The two beds were divans.They were good wide springy divans with useful drawers underneath, but there were no bedsteads and no bedposts; nowhere to hang up a stocking.

Granny overheard Marnie and came in.

'Don't worry about that,' she said. 'We're going to do what I used to do when I was little,' and on Christmas Eve they found out what she meant. Granny's fireplace had a wooden mantelpiece and she hung a string from one side to the other like a little washing line, with two clothes pegs clipped to it.

'Much easier for Santa Claus,' Mum said, when she saw Marnie looking doubtful.

Much easier for Mum, Stephanie thought. She won't have to tiptoe about and I won't have to pretend to be asleep.

'I'll get our stockings,' Marnie said. They were not really stockings. They were the thick woolly socks Mum wore to work, to keep her feet warm on the chilly station. She was the person who collected tickets.

'No need,' Granny said. 'I've been knitting,' and she gave them each a new stocking to hang up, long proper stockings, proper Christmas stockings, in red and white stripes.

When Stephanie went to bed she took a last look at them, hanging there together from the line above the fireplace, at the holly along the mantelpiece, the glittering tree in the corner, Mum and Granny enjoying a glass of beer beside the dying fire.

'Sleep well, darling,' Granny said.

'Don't wake Marnie,' Mum said, and at the same moment they

all heard Marnie yell, 'It's snowing!'

She came running downstairs and they went to the front door to look.

'This is the first white Christmas for twenty years,' Granny said, and Stephanie, as she climbed the stairs with Marnie, thought that almost she could believe again in sleighbells and reindeer and Santa Claus.

She woke at midnight. She could see the luminous hands on the bedside clock, so close together that they looked like one, and down below, the clock in the hall was striking. But it was not the sound of the clock that had woken her; something else had made a noise, a furtive bumping and fumbling. Stephanie's first thought was, Mum, being Santa Claus; then almost at once she recalled that this year the stockings were downstairs and when she peered over the hump of her shoulder beneath the duvet she saw that the room was empty.

It was easy to see because the snow reflected Granny's porch light and made the room glow palely. Perhaps Mum had just looked in at them on her way to bed. Stephanie was about to turn over and fall happily asleep when she heard the sound again, very close and rather louder. It seemed to be coming from behind the wall between the beds.

Marnie was sleeping deeply, curled into a ball with her head under the duvet. Stephanie silently rose up in bed and placed her ear against the wall. It felt warm, and she had just time to remember that beyond it lay the chimney of the living-room fireplace, when she jumped. The sound was definitely coming from behind the wall, a muffled scrabbling, lumping, rasping; the sound of something large trying to work its way through somewhere small, the sound, in fact, of someone coming down the chimney.

If it had been any other night of the year Stephanie would have swarmed back to bed and disappeared under the duvet like Marnie, but although she was afraid, she was not scared out of her wits, and her wits told her that there was only one person who could possibly be in a chimney at midnight on Christmas Eve.

She didn't believe it, but she *knew*. Barefoot and on her toes she crept across the room, opened the door and slipped out on to the landing. Behind her, strange sounds continued in the chimney, but lower down, now, nearer to the floor.

There was no need to go all the way to the foot of the stairs. The living-room door had been left open and Stephanie, peering through the banisters, could see the fireplace. Granny must have opened the curtains before she went to bed, for the soft snowy light shone in. In one corner Stephanie could just make out the dark bulk of the tree, a pile of presents at its foot. There stood the chairs where Granny and Mum had sat, and there was the grate, full of cinders, the hearth swept clean and ready for tomorrow. There were the stockings, full and fat and lumpy, and there, in the fireplace, just coming into view above the empty grate, were two fat and lumpy feet.

Forgetting to breathe, too astonished to move, Stephanie crouched on the stairs and watched as the feet danced uncertainly in the air until they touched the bars of the grate, and then vanished as something like a heavy cloak fell around them. From among the folds of the cloak an arm emerged, and then another, groping and groping until the hands at the ends of the arms discovered the dangling stockings; and tugged. There were two sharp clicks as the pegs flew off, the washing line twanged and the dark shape shrank back into the fireplace.

Stephanie stayed on the stairs only long enough to see that the stockings were no longer there before hurtling back to the bedroom and diving into bed, deep, deep under the duvet, but not so deep that she could not hear, from behind the wall between the beds, the bumps and thumps and mumbles of something going back *up* the chimney.

In the morning the snow lay bright and thick. Stephanie let one ear emerge from under the duvet to hear Mum come in and say, 'I hope you haven't got soot on the sheets.'

'What soot?' Marnie said.

'Some soot came down the chimney during the night. Did you tread in it?'

'I just woke up,' Marnie said.

'Then where are the stockings?' said Mum.

'I'm just going to get them,' Marnie said, and ran out of the room.

'But they aren't there,' Mum said.

Stephanie came right out from under the duvet and saw Mum looking at her.

'Where are they?' Mum said.

'Where's what?' said Stephanie, but she remembered last night and knew that what she had hoped was a dream was not.

'The stockings,' said Mum. 'Did you go down early to get them? I thought I heard someone creaking about. You might have let Marnie fetch her own.'

'I haven't got them,' Stephanie said, as Marnie appeared in the doorway holding out two clothes pegs and ready to weep. Her feet were covered in soot now and so, when they went down to investigate, was the living-room fireplace. Soot lay on the tiles as thick as snow upon the lawn, and there were large, shapeless prints in it, much larger than Marnie's little feet could have made.

Granny and Mum checked all the doors and windows.

'Burglars?' they asked each other, but nothing else was missing and there were no footprints in the snow outside.

To make up for the lost stockings Stephanie and Marnie each had a present from the tree before breakfast, while Mum muttered something about sleepwalking and went up to frisk the bedroom.

Granny swept away the soot and lit the fire. Stephanie helped and said nothing.

It was only after breakfast when all the presents were opened that they thought of going out to play in the snow. The sun was shining now. Marnie began to roll a huge snowball. Stephanie went to the end of the garden and looked up at the roof. Something had caught her eye, something bright and out-of-place. The snow in the garden was already churned up but the snow on the roof was smooth and pure; except around the chimney stack. There it was scuffed and sooty, littered with scraps of torn gift wrap and sweet papers and ringlets of orange peel. Even from so far away Stephanie could tell that it was orange peel. She also recognized what looked like two red and white streamers draped over the chimney pot.

She did not say anything to Marnie, who was busy with her snowball, but she fetched out Granny and Mum and stood with them at the end of the garden, staring at the roof and waiting for one of them to come up with a sensible explanation which, after all, is what grown-ups are for.

In the end, after a long silence, Mum said, 'It must have been Santa's no-good brother, following him round. One fills stockings and the other nicks them.'

This was not, of course, a sensible explanation, but nobody could think of a better one.

Pteranodon

Douglas Livingstone

A seven year old herd boy,
ragged happy and vacant,
sits alone playing the stonegame,
his back to the five
thin healthy head grazing.

Across the valley
the distant warts of huts
squat on the wrist of the hill.

Long believed extinct,
there was no one
but the wall-eyed
stampeding clot of cattle
to see the two dozen
feet of dusty leather
wingspread, hear the wet
crush of long toothed jaws closing,
the snap of vertebrae,
and nothing, nothing at all
the flight away
with the broken rabbit boy
one limb slow waving.

The Leopard in the Rafters

Marilyn Watts

This is a true story, my grandmother told me. It didn't happen in our village, though—it was in a smaller group of houses, a bit further down the mountain. The story was told to my grandmother by her aunt. And it happened a long time ago. It must have been a long time ago, because no one we know has seen a leopard around here in years.

Leopards are very powerful. And secretive. Out in the bush there are lions, and you can hear them grunting as they prowl at night. But leopards are silent, invisible. They can live near villages without anyone seeing them—and then a goat will go missing. I used to have nightmares about leopards. The thought of them has always terrified me. Perhaps, I used to think when I lay awake, there's one living near by that we haven't seen . . . and I wouldn't go outside our house at night.

Grandmother tells us stories sometimes, after the evening meal. We put out the lamps, to save paraffin, and listen by the glow from the cooking fire. It's very dark outside. It's very dark in the corners of the room, and in the blackness of the roof above us, anything could be hiding.

You know where the road bend round, to Mwanga? Grandmother asks, sitting back on her heels and drawing in the earth on the floor. And the road go along and past a place all growed over? There was a house there once. You still see the walls behind the old banana trees. Only no one don't live there now. Not after it happened.

And we imagine the road and remember the ruin, and we grip our knees and listen.

Old woman, she live there in that house with her husband. The banana trees are all that is left of their house and the shamba. They grow all their food on the shamba, hardly ever go to market. So long a walk. The husband was a good man, but the mama was greedy. Too greedy. That's why it happened.

One day they have some money, so the old woman she go to market. All the way there, she think about what she want to buy, and what she want to eat. Meat. They hadn't had meat in so long. So she walk to the market with a young girl from the village, and she think about the meat. Goat? Or beef?

While she go, there's a leopard. It's an old leopard, cunning, and hungry. There's no one at home in the hut, and the mud of the old walls is crumbled away in one place so the cat climb in. Careful. Quiet. Sssh. He smell man, but there is no one there.

The mama walk along, she's thinking about meat. Her husband is away on the other side of the village, helping make the mud bricks for a new hut. In the silent, empty house there's the smell of food. So the leopard climb in and lie down along a rafter. And he wait.

We imagine the leopard lying there, still. Its legs and tail hang over the rafter, just as they sleep along the branches of the trees during the hot afternoons.

Old woman she buy food, she buy the meat, and they walk back along the road. The girl is hungry, and she hasn't had meat in many days, but she know she won't get none. The old woman is greedy and also selfish—everyone know that. So the girl, she sigh and say goodbye at the old couple's shamba, and walk on. She walk through the village, past where the husband is working, but she don't tell him his mama is back with the meat. He can wait, she think.

The old woman go into the house, mouth watering. She open up the parcel of meat and look at it. And then she think. If she cook it now, and eat it now, her husband will never know. She can keep a little bit of meat for his supper. And she can eat the rest herself. She lick her lips at the thought of it—mouthfuls of hot, juicy meat.

In the roof darkness, between the beams and fire-black banana thatch, the leopard lick his lips. And once, just once, he flick his tail.

We shiver, and move a bit closer to the fire. Daudi grins at me, a little nervous, but this is his favourite bit.

The old woman start the cooking fire burning, and fetch a cooking pot. Then she close the door. She shut it tight. It's very dark in the house, but now no one will come to call. And no one will smell the meat cook.

But the leopard he smell the meat as soon as she open the parcel. He don't need it cooked. And he smell other meat, too. The leopard open his golden eyes very wide.

The woman she put the meat on the fire and sit quiet, crafty, watching the juices run. She swallow and lick her lips again. Above, in the rafters, the leopard lick his lips. And quietly he stretch, balance on the wood, and jump down—claws sharp, reaching for her back, jaws open.

The door was shut, you know, and no one heard her scream. She make sure her husband and the girl were away, so no one come running to help her.

Grandmother told us the story for a reason, of course. All her stories had a moral. Afterwards she would say: so you see, you must never be greedy. But that's not what I remember. The story had a different effect on me.

Since then, I'm careful every time I go into someone's house. I make sure I sneak a look, just a very quick glance, up at the ceiling. Just in case there is something lying there in the shadows, along the rafters. . .

The Friendly Ghost

A C Bolton

Tears were streaming down the face of Lal Das, and he was sobbing so violently that it was a minute or so before his eleven-year-old brother, Ram Das, could get anything out of him.

'Is it a tiger?' Ram asked patiently. 'Has something happened to our father? Has the moneylender asked for more rupees? Come, little Lal, tell me.'

Lal finally controlled his sobbing, and the tale he told brought horror and fear into his older brother's eyes. Lal's work was to watch over the new water buffalo which all the village had saved and scraped for over a period of months. With this magnificent beast they would be able to plough more paddy fields for rice. If the gods were kind, and the monsoon rains came at the right time, there might be plenty of food in the tiny Bengal village when harvest came. Now it seemed as if all that was a thing of the past.

The big grey water buffalo had insisted on walking into a stretch of land which was haunted by a ghost. For years neither man, woman nor child had walked there. The land was accursed, for it had swallowed up two children and a man four years before.

Each time one or more of the villagers had thought they heard a sound like the distant clap of thunder, and each time a victim had vanished leaving no trace to tell how they had gone. With a rusty roll of barbed wire, left behind by the soldiers who had fought the Japanese, the villagers had wired off the haunted stretch of paddy field. Now, with the wire more or less rusted away, Lal had been unable to stop the new buffalo from walking on to the forbidden territory.

Ram Das left his work, and walked quickly to the forbidden ground. Once it had been fruitful paddy land. Now it was deep in grass, with scrub bushes beginning to turn it into wilderness again. The jungle was slowly reclaiming its own.

The two Bengali children stood and looked across to where the buffalo grazed. For the moment, at least, nothing had happened. In fact, the scene was as peaceful as it always seemed to be. A kite which had been wheeling in the hot air, suddenly swooped. It vanished into the long grass, and came up carrying a wriggling Russel viper in its beak. A hundred feet into the air it went, on powerfully beating wings; then the viper was allowed to fall. The kite swooped after it, picked up the dazed reptile again, and repeated the process until the viper was hurt so badly that the kite was able to finish it off, and carry it away to be eaten at leisure.

'Three times the kite swooped there,' Ram Das said soberly, 'yet it has flown away without injury. Perhaps I could bring back the buffalo without . . .'

'Nay, nay,' his younger brother protested. 'Is it not forbidden to go there? If the ghost is angered . . . you, too, may disappear.'

Ram Das nodded. He knew that; yet he knew also how much the buffalo meant to his father, mother, and all the others in the poverty-stricken village. He tried to imagine his father's horror when he heard what had happened to the buffalo, and that decided him.

'Tell our father,' he ordered. 'Perhaps he will know what to do. I will stay here and watch.'

Lal Das did not need a second bidding, but shot off like an arrow, his bare feet pattering on the sun-baked ground, his soiled turban in his right hand.

As soon as he was gone Ram Das stepped fearfully over the rusted barbed wire which marked the beginning of the ghost-ridden land, and began to walk towards the big buffalo.

The heavy great beast, bulkier than any prize British bull, seemed to realize that the lush grass on which he now fed was soon to be denied him, for he snorted, and began to walk further away from young Ram Das. The fringe of the jungle smelled cool in his nostrils, and the sun beat down with pre-monsoon fierceness.

The young Bengali shouted to the buffalo, hoping to stop him. He only shouted once, suddenly remembering that if the ghost heard him he might vanish as the others had vanished.

His heart was thumping madly as he passed the first of the clumps of scrub. They had an ominous look, as if they could quite easily hide a hungry ghost.

Yet nothing happened. The buffalo stalked onwards, snorting now and then as the tall grasses tickled his nostrils. Ram Das was whimpering a little in terror when he finally reached the smooth-hided beast.

'You fool,' he said angrily. 'Have my father and the village elders starved for a year so that you can come here to be snatched away by a ghost. Back . . . back with you,' and small, thin lad though he was, he struck the buffalo shrewdly across the nose.

With one sideways sweep of his great curving horns the buffalo could have sent young Ram flying, but instead he tossed his head, grunted, and turned back.

As he did so, the grasses a yard or so behind parted, and the grinning mask of a tiger showed, its stripes blending almost perfectly with the dappled light and shade.

Boy and buffalo saw the movement, however. They both stopped. For a few moments the trio remained motionless; the tiger staring at the boy and the beast, the other two staring at the two yellow orbs in that grinning mask of a face.

Then the tiger sank slowly down. His tail, which had been moving from side to side, twitching the grasses, became still. Ram Das knew what that meant, for he had seen tigers before. Most tigers, he knew, will not molest man unless they are injured, or driven at bay by hunters. Occasionally, however, when a tiger is growing old, and finds difficulty in killing, or if injury has baulked him, then he will attack humans.

This tiger was lean flanked. He limped from an injury in a front paw, and his hunger was like a gnawing terror which drove all fear of human beings away. He crouched for the spring which would bring him food, and young Ram Das knew that unless he did something quickly he was as good as dead.

This, he thought, is the ghost! This is the thing which took the others. His heart almost bursting with terror, he shrieked desperately.

For the moment that shriek saved him. The tiger, in the very act of springing, checked, and snarled. The sound broke the paralysing fear which had held the buffalo motionless. The beast turned and fled at a lumbering trot.

Ram Das turned to run, but checked himself in time. He knew that once he turned his back on the tiger, the beast would be on him. While the buffalo trotted in ungainly fashion towards the village, the young Bengali began a backwards retreat, never taking his eyes from the yellow orbs which were almost hypnotizing him.

The tiger came out of the taller grasses, slinking forward slowly, almost hesitantly. That shriek had shaken his nerve, yet with each stride he took, the hunger drove him nearer a kill. He could tell the boy was afraid. There is something in the sweat of fear which an animal can sense, and beads of perspiration were rolling down Ram Das's face.

Somewhere in the direction of the village there was a growing clamour. Young Lal had roused the villagers, and they were coming. But they did not know of the tiger. The precious buffalo was well on the way to being safe, but Ram Das was nearer death every second.

He quivered with renewed agonizing fear as he backed into a bush, and his brown skin prickled with blood in a dozen places before he freed himself. He whipped off his turban and threw it at the tiger. The beast swept up a paw and snarled, yet it still hesitated. The dark eyes of Ram, staring into its own, held it at bay.

Then, from the edge of the forbidden land, came a wail of horror, half of entreaty. It was the father of Ram Das, pleading with his son not to run. They were sending for Ali, Ali Chatteranjan, who had been a soldier. Ali had a gun. If Ram could hold the tiger long enough then Ali would shoot it.

Ram Das was almost beyond listening. In his terrified eyes the tiger seemed to be growing larger and larger. For a few moments, spurred on by his father's pleading, he did straighten himself, and walked steadily backwards, waving his arms, and shouting at the tiger; by now, however, the striped killer seemed to have realized that more humans were drawing near. He ran forward a couple of paces, dropped belly to earth, and would have sprung there and then if Ram had not turned and fled, yelling with fear.

He did not know where he was going. All he wanted to do was

to run, run as he had never run before. He leapt over a small bush which stood directly in his path, and the tiger, snarling, followed in hot pursuit.

Had he not been hampered by his lame foreleg he would have caught the young Bengali in a couple of bounds, but that weak forepaw tripped him. Ram Das turned at a big bush, the tiger swerved to head him off. Two mighty bounds, and he was near enough for a spring. He dropped, belly to earth, gathered his powerful muscles, and seemed to float through the air—an eight-

foot length of tawny striped body, long tail behind him like a plume, right paw swinging out slightly for the blow which would smash the back of the boy.

Ram Das saw the tiger in the air out of the corner of his eye, and he 'jinked' as he had seen wild pigs 'jink' when pursued by mounted spearsmen. He thrust his right foot forward, put all his strength behind the thrust, and hurled himself to the left. He shot to one side and, missing his foothold, fell heavily on the ground.

The wail from the watchers—half the village had come out to see what was happening—was drowned by a new sound as the tiger hit the sun-baked earth. The forepaws touched earth three yards from Ram Das, and an instant later there was a sudden roar, like thunder. The ground blossomed smoke, earth, and the sprawling body of the tiger.

Then everything was hidden by the billowing smoke and for a few moments, as the smoke went upwards into the drowsy air, the watchers ducked and squealed as tiny pieces of metal, stones and clods of earth, showered about them.

The ghost had spoken again.

When the crash of the explosion had died away, and the ears of the terrified villagers were normal once more, they heard a strange sound from the 'ghost' land. The terrified sobbing of a boy.

Young Lal Das, forgetting his fear, ran to where his brother was lying face down. Yards away was a strange hole in the ground. Still further away the mangled body of a tiger twitched in death.

Save for one or two scratches Ram Das was uninjured. The buffalo was uninjured. The ghost, which had killed two children and a man years before, had spoken again. Only this time it had saved a boy's life and killed a tiger.

A week later three soldiers came with a strange thing on a stick. By its aid they were able to tell where the rest of the landmines were buried.

Landmines, the soldiers said; but to Ram Das, and the others in that village, the thing which killed the tiger was a friendly ghost . . . and that stretch of paddy is still untilled.

Kiitos! Kiitos!

Rachel Hands

Cindy padded over to look at her mother's little travelling clock. 'Mum! Dad! Wake up! It's late. We've overslept!'

'Too early,' mumbled her mother. 'Go back to sleep.'

'But it's morning. It's gone breakfast-time, and I'm hungry.'

'Get a biscuit and go back to bed,' said Mum, and rolled over without opening her eyes.

Go back to bed? On the first day of the holiday? A soft golden day like this? Cindy explored into the kitchen, and found a box of juice and a packet of biscuits left over from the journey. She looked back at the bedroom, considering. Dad was tired after all that driving, and Mum was always sleepy after her travel pill. It seemed a pity to disturb them again. She dressed, took a left-over apple and her own bar of airport chocolate, wrote *I have gone to the lake* on one of Mum's lists, and let herself out.

The track wound through the trees, but Cindy could see a gleam of water below, and took a short cut to arrive with a jump almost on to the road at the edge of the lake.

There was no one there.

'Makes a change,' she thought, remembering the noise and crowds of last year's holiday village in. . . Wales, was it? She walked along to the shop but it was shut. 'Closed early for lunch,' she decided, peering inside, although it was barely twelve by the clock above the counter.

Lunch? It was breakfast she wanted, and she had no money. Still, she could play with the other children till Dad came to fetch her. She ran up the slope to the café.

There was no one there. The door was locked, and the big room was bare, chairs and tables stacked at one end. Through the kitchen windows she could see a single mug, upside down, on the draining-board but no kettles, no pots and pans, no tea-cloths. Outside, the dustbins stood empty.

'How queer.'

Cindy wandered back to the lake and stood looking at the little jetty and the landing-place cut in the reeds. The water gleamed softly in the pearly light that shimmered through a pale cloudy whiteness. The sun had gone in, gold and orange-pink giving way to blues and mauves and greys. Only the log-cabin café glowed deep red-brown, the trees shining palely beside it. Behind, the bare grey boulders of the fell reared up into the sky.

It was very still, and Cindy was quite alone. 'It would be beautiful if it weren't so spooky,' she thought.

Spooky. Her insides seemed to jump and huddle together. Something started pushing up inside her throat. It was colder, and the queer light was suddenly queerer. In the silence she could hear her own breathing. Suddenly she turned and started to run back up the track, ducking her head to keep from seeing those tall, pale trees.

'Mum, I don't like it. Please, Mum. Please, Mum, please come. I don't like it. Please.'

Out of breath, she stopped. All the chalets she had passed were shut up and empty. Their own, with the hired car outside, was nowhere to be seen. She broke away through the trees to find another track. 'Please, Mum. Please come.' And then, all at once, she was lost. There were no chalets, and no tracks, only herself and the trees, and the eerie light.

'They'll come and find me,' she heard herself saying. 'I didn't just go out. I left a note. They'll come and look for me.'

But down at the lake, not here. And there was no way back to the lake now. The ground was level, and no gleam of water showed through the trees. She began to shiver, and spoke firmly to herself in her mother's voice: 'If you are lost, go to a shop, or a house where you can see there's a family with children, and ask for help.' But the shop was down by the lake, and was closed for lunch, and the café was deserted, and all the chalets were empty. There were no people in this place except herself.

She found a tiny path and followed it, till at last she came to a chalet with a car outside. Not their own, but there were children's toys on the back seat and in the porch, and 'This is a family house,' she thought. Suddenly shy, she knocked gently. What if they didn't talk English? Dad had said that before you went to someone's country you should learn a bit of the language, but all he had taught her was *kiitos*, thank you. She tried it softly now, but there was no need. No one came to the door. Timidly she knocked again. No one answered.

'They're all out,' she thought, beginning to shiver again, but in her head she heard her mother's voice: 'If there's no one to ask, stop rushing around. Find a place to wait, and *stay there* till someone comes for you.' It was not a comfortable place to wait, but she settled into a corner of the porch, ate half her chocolate, and sat staring straight ahead into the silence.

It was much, much later that she heard the noises. A swishing as the undergrowth was disturbed, and a soft *click* . . . *click* which brought the tight feeling back into her throat. After a while the swishing stopped, but the clicking grew louder and more regular. Something—some *things*—were coming along the track. Cindy pressed back into her corner, hands over her eyes, thumbs reaching back to block her ears. Then a greater fear, of being caught unawares, made her peer cautiously through spread fingers. Whatever it was—whatever *they* were—had left the track and gone pushing away between the trees. She caught a glimpse of big swaying, shaggy bodies—brownish, greyish, and the last a pure ghostly white; she heard again that soft eerie clicking; then all was quiet.

She sat very still for a long, long time. Somebody *must* come now. The sun had come out again, and 'It must be nearly tea-time,' she thought. 'I'll finish my chocolate and eat my apple, and then I'll walk right round the chalet, and then they'll be here.' She took it very slowly on purpose, fingering the rough logs of the walls and peeping into every window. At the back were the bedrooms: in one were three children, in the other their parents.

They were all fast asleep.

Cindy left the chalet at a run. She knew now why her parents hadn't come. They were asleep in their chalet, just as these people slept in theirs, undisturbed, unwaking—for how long? 'And I think I was falling asleep too,' she shuddered, 'only those creatures came for me.' She must get away from the trees. 'Someone must help me, somewhere,' she panted as she ran. 'I don't *want* to be the only one left.'

'Kiitos,' she found herself saying, over and over again. 'Kiitos, kiitos'—only now it sounded like a *please*, not a thank you. She took the path the strange creatures had shown her. Were they magic beasts, come to lead her to safety, or were they more dangerous even than the trees and the queer light and the log-cabin chalets?

Where the forest ended, a stretch of rough stony grass led across to the first slopes of the rocky fell. 'If I could get up there,' she thought, 'just a little way, perhaps I could see what to do.'

But between her and the fell stood the strange creatures, silently watching. Away from the trees, in the sunshine, they didn't *look* magic . . . but a sudden movement made her jump, and a banging start in her ears. What had seemed just a fallen log had scrambled up and lumbered away, huge horned head held low, clumsy feet grotesque beneath thin bony legs. Beyond, another rose up, and another, and another, and sharply over the still air came that soft insistent clicking.

'I won't look any more,' she told herself. 'I'll keep to the path and go straight up the mountain. If they come for me, perhaps they won't like to go on the stones.'

Looking straight ahead, trying not to run, she made for the nearest part of the slope. Where the grass gave way to the first grey boulders, she hurried, climbing, scrambling, slipping, until she was as high as she dared to go on that stony mountainside. Then, cautiously, she stood upright and looked all round. Behind her, the strange beasts grazed on the tussocky grass, almost invisible now among the rocks and fallen trees; before her, far below, lay the road, the café, the shop, and the lake. By the edge a man stood watching three small boys as they played on the jetty.

Going down was harder than going up. She stumbled and slithered, afraid to take her eyes off the group by the lake, but she reached the path and began to run to where they waited, watching.

'Please,' she panted, 'please—' but the thing in her throat was too much for her. One of the little boys said something to her, and another took her hand.

'They want to know why you are crying,' said their father.

Shaking all over, Cindy tried again. When she had finished, the man spoke.

'It is O.K.,' he said. 'First, your father and mother have not yet come, I think. But soon they will wake. They do not sleep for a hundred years. Second, no one is here, because it is a ski-resort. In winter, in the snow, all is lights and noise and crowds. Now it is summer, and only few people come, for the loneliness and the quiet. Third, those animals are not magic beasts. They are reindeer. They live here.'

'Reindeer? *Those?*' Cindy was startled. She thought of Christmas cards, and shook her head. Then, 'But just *loose*? Just walking about? And they *click*.'

'That is their feet, their hoofs, the way they are made,' said the man, and took a package from his pocket. 'We will go and find your father and mother,' he said, 'but first, I think you are hungry.' He gave her thin slices of cheese and a lump of hard dark bread, sharing the rest with the three boys. 'For breakfast,' he said.

'But it's tea-time.'

'No.' He smiled. 'You have been walking all night, not all day. Here in Lapland, in the summer, the sun never sets, and we have no dark. Did you not know?'

'Oh . . .' Suddenly Cindy understood. '"The Land of the Midnight Sun." But I didn't think it would be like *this*. I thought it would be ordinary proper sun—like day-time. Not so strange, not like this.'

'We think it is beautiful,' said the man.

'So did I,' Cindy remembered, 'before it was spooky. It *was* beautiful,' and she looked back across the lake. 'It *is* beautiful. I'm sorry.'

The man smiled again. He turned her round to see someone strolling down the track towards them.

'*Daddy!*' Cindy was off like a rocket. Then she braked hard. She stopped and ran back to the little family at the jetty.

'*Kiitos*,' she said. 'Oh, kiitos, kiitos!'

Empty Fears

Brian Lee

What's that?—Coming after me, down the street,
With the sound of somebody dragging one foot
Behind him, who pauses, who watches, who goes
With a shuffle and mutter
From the wall to the gutter
In the patch where the light from the lamps doesn't meet . . .

Oh . . . it's only a bit of paper—a hollow brown bag
Open-mouthed, like a shout—a bit like the face
Crumpled-up, of someone who's going to cry,
Blown on the wind, from place to place,
Pointless, and light, and dry.

Who's that?—Watching, from the upstairs windows
Of the house where the hedge grows right back to the door,
Where the half-drawn curtains droop and discolour
And a yellow bulb burns away
And the milk's on the step all day—
Somebody lives there, no one comes or goes . . .

Oh . . . it's only an empty coat on a hanger
That sways in a draught like a man who depends
On only one thing—the something inside
That's holding him up, waiting for friends
He writes to, but no one's replied.

What's that?—Whispering, where the fence round the lot
Sags like a fading hope: the gate just here twists
On its hinge like a bird's broken wing
And shrieks as you look, and see:
Nothing, where all the shops used to be,
People coming and going where now they are not . . .

Oh . . . it's only the breeze, that's fretting itself
Amongst the stiff thistles, each standing alone,
Upright, all winter, dead, but not gone . . .

But if it's only these things, what blows
Through me, to make me afraid, who knows?

Exit

Patricia Miles

'Come on,' said Hawkins eagerly, in his piping voice. 'Let's get off at the Post Office. We can soon walk up there.'

'In this?'

Carter gazed out doubtfully, as the bus rocketed along the empty road. Every now and then a few fat drops of rain fell out of a yellowish sky and dashed against the window. Beyond the hedges the fields, caught in the strange light, showed a living, incandescent green.

'Go on, be a sport. Come with me, will you? This is practically *ideal*.'

It wasn't the sort of weather most people would have called ideal. The storm had been brewing all day, with thunder rolling in the distance. What Hawkins meant was 'ideal for him', i.e. ideal for trying out his theory.

'Come on,' said Hawkins again, getting agitated, 'we're nearly there.'

'Oh, all right,' Carter said then, out of good nature.

They grabbed up their school bags and Hawkins rang the bell. The bus stop served a few cottages and a farm a little way up the hill. One of the cottages was also a Post Office. Opposite the Post Office a track led up a gentle slope to a ruined chapel—their destination.

They crossed the road—there was hardly ever any traffic—and set off up the hill, rather clumsy and heavy-footed on the stones of the muddy path. They were laughing and larking about as they walked along.

'What exactly are you hoping to see?' said Carter. 'If hoping is the right word.'

'I'm not sure. Monks, white ladies, something like that.'

'What about the old nameless dread, eh?'

'That too.'

The path narrowed and they got into single file. They hadn't far to go. The chapel was only a field or two away from the road, in a small wood fenced off—inadequately—with barbed wire. You could actually see the decayed end wall from the bus, if you looked at the right moment. All the same, close to civilization though it was, it had somehow succeeded in keeping its own peculiar atmosphere.

To his surprise, Carter found his mood changing: all his jokiness was dropping away: a feeling of oppression sat heavily in his chest. He wasn't going to let on to Hawkins, though. He decided it must be the weather. It was the queerest weather, so heavy and overcast. Thunder rumbled again. It was getting nearer.

'Listen to that!' continued his friend, still in the same loud tones. 'We should see something. Or it could be we'll just feel a drop in the temperature.'

Carter wished he'd let him come on his own. There wasn't all that much of a friendship between them: they just travelled in the same direction from school, and usually caught the same bus home. They weren't even in the same year. Hawkins was a clever little squirt out of Form Two, all specs and long words. Carter was older and brawnier. When they sat on the bus together Hawkins liked to rattle on, and Carter, who was good-natured, let him. He wasn't much of a talker himself. Hawkins was always full of ideas.

'Tell me it again,' said Carter, out of a dryish throat, 'your idea.'

They were nearly at the wood.

'It's simple, really,' piped Hawkins, importantly. It wasn't simple at all, except at the start. 'My theory is, there is some connection between thunderstorms and psychic phenomena. You know that photograph I showed you out of the library—that one with the ghostly shape on it? That was taken at Corfe Castle just before a thunderstorm.'

'Bet it was a fake.'

'Maybe,' said Hawkins with scientific open-mindedness.

'And anyway—it wasn't taken here.'

'No, but it was in this sort of place, a well-known spooky spot. Even if that one was a fake, it doesn't really matter. The point is, there are so many accounts of apparitions and hauntings connected with storms, there must be something in it.'

'I wouldn't have thought you believed in ghosts.'

Hawkins shrugged. 'Well, I don't believe in ghosts, exactly. What I really think is, *either* something in the past leaves a sort of photograph, or film of itself, quite by accident, which you can pick up when the atmosphere's right; *or*—this is my other theory'—he paused impressively—'beings *from other worlds* reach through to us here. Whichever it is, it's connected with these special places, and storms.'

'*Other worlds?*'

'Yes. You know, out of deep space, or another time—what do you call it—an alternative universe. You see, if people here happened to see something like that, they'd naturally interpret it as a ghost, or the devil, or something supernatural, wouldn't they?'

Carter stood still, apparently to get his breath. 'You mean *we* might see things from outer space.'

'Lots of people have—UFOs and all that—they're always seeing them in Wales. Or we might see a funny sort of historical film, if my first theory's correct. Boy! If there is anything here, this is the day to see it.' He hurried eagerly up the last few yards of the path. Carter followed, slowly.

'Hawkins, what are you *supposed* to see here?'

'Dunno. I just know that it has this reputation for being haunted.'

It was easy getting into the wood. The wire had rusted through in places. There was a strong scent of bluebells everywhere; they were almost over, dying mostly. You couldn't walk without treading on them.

'Nothing stirring yet,' said Hawkins. In the wood even he sounded subdued, as if he had suddenly realized what in fact he might see. Carter moved his shoulders uneasily: it was a curious animal uneasiness, purely physical. He had not believed a word Hawkins had said.

They went on towards the chapel. Most of its grey stones had been carted away years ago, but the foundations and one wall remained. The thunder sounded again, but it seemed further off now.

'Well, this is it.' There was a quaver in Hawkins's voice, and he jumped when a spatter of rain fell on him. A raindrop ran like a tear down his glasses. And then he disappeared.

Suddenly, and with great completeness, Hawkins was no longer there. Carter had been staring right at him, even noticing the yellow streak of light reflected in his glasses and the splosh of rain. The light had touched the wall too, just above his friend's head.

Wait. It was at this point that the blood in Carter's veins chilled to ice water, and that each hair rose separately on his head and arms.

There was *no wall*. He turned round. There were no trees, no electric pylons, no cottages, no road. The lie of the land was not the same. There was no storm. The sky was clear, a curious pale mauve colour. There was no sun. He was standing alone in a rocky place, and a small wind was blowing. A wild facetiousness swept over him. He started to laugh. *He* was the one who had disappeared. Exit. Finis. Out goes he.

Then he burst into tears. 'That Hawkins, when I get hold of him, I'll kill him. I'll duff him up. I'll make him wish he'd never been born.'

Of course, he was never able to do any of these things.

ACKNOWLEDGEMENTS

The editor and publishers are grateful for permission to include the following stories in this collection.

Dannie Abse, 'Emperors of the Island' from *Collected Poems* (Hutchinson). Reprinted by permission of Sheil Land Associates Ltd.

Vivien Alcock, 'A Change of Aunts' from *Ghostly Companions* (Methuen, 1984). Reprinted by permission of Methuen Children's Books and John Johnson Ltd.

Daniel Wynn Barber, 'Tiger in the Snow' from *The Year's Best Horror Stories XIII*, ed. Karl Edward Wagner (DAW Books, 1985). © 1984 Daniel Wynn Barber. Reprinted by permission of the author.

A C Bolton, 'The Friendly Ghost' from *Super Books of Ghost Stories*, ed. Leonard J. Matthews (Hamlyn, 1977).

Ruskin Bond, 'Eyes of the Cat' from *Time Stops at Shamli and Other Stories*. Reprinted by permission of the author and the publishers, Penguin Books India Pvt. Ltd.

Dave Calder, 'This is the Key to the Castle' from *Bamboozled* (Other, 1987). © Dave Calder 1987. Reprinted by permission of the author.

Charles Causley, 'Miller's End' from *Collected Poems* (Macmillan, 1975). Reprinted by permission of David Higham Associates Ltd.

Kevin Crossley-Holland, 'Slam and the Ghosts' from *British Folk Tales* (Orchard Books, 1987). Reprinted by permission of the publisher. This is Crossley-Holland's retelling of a Scottish tale collected by Kenneth Goldstein for the School of Scottish Studies. The original story is in Katherine M Briggs: *A Dictionary of British Folktales* Part B Volume 1 (Routledge and Kegan Paul, 1971).

Hugh Sykes Davies, 'Poem'. © Hugh Sykes Davies.

Roy Fuller, 'The Start of a Memorable Holiday' from *Seen Grandpa Lately* (Scholastic Children's Books). © The Estate of Roy Fuller.

Adele Geras, 'Camilla'. © Adele Geras 1991. Reprinted by permission of Laura Cecil, Literary Agent.

John Gordon, 'The Hawk'. © John Gordon 1985. Reprinted by kind permission of the author and A P Watt Ltd.

Grace Hallworth, 'The Shiner' from *Mouth Open, Story Jump Out* (Methuen, 1984). © 1984 Grace Hallworth. Reprinted by permission of Methuen Children's Books. In this collection of scary stories from Trinidad, Grace Hallworth says of 'The Shiner': 'La Diablesse (Lajables), or lady-devil, is a witch who usually travels by night . . . She may lead [men] to their death unless they are lucky enough to spot the cloven hoof beneath her long full skirt . . .'

Dennis Hamley, 'Supermarket'. © 1992 Dennis Hamley. Reprinted by permission of the author.

Rachel Hands, 'Kiitos! Kiitos!'. © 1992 Rachel Hands. Reprinted by permission of the author.

Gregory Harrison, 'The Cave', first published in *A Fifth Poetry Book*. © 1985 Gregory Harrison. Reprinted by permission of the author.

Virginia Haviland, 'Wiley and the Hairy Man'. This has been slightly edited by Virginia Haviland from a story collected by Donnell Van de Voort. Van de Voort's story is in B A Botkin (ed): *A Treasury of American Folklore* (Crown Publishers, 1944), but there are several different versions. The Hairy Man, who 'didn't have feet like a man but like a cow', is the devil himself of course.

James Kirkup, 'Who's That?'. Reprinted by permission of the author.

Sheila Lavelle, 'Dear Jane . . .'. © 1992 Sheila Lavelle. Reprinted by permission of the author.

Maria Leach, 'The Big Black Umbrella' from *The Rainbow Book of American Folk Tales and Legends* (World Publ. Co., 1958). This retelling is based on John Bennett's story 'The Remember Service' in *Doctor to the Dead* by John Bennett (New York: Rinehart & Co. Inc. 1947). The story was told to Bennett by Mary Simmons herself, who showed him the umbrella. There was a tradition among South Carolina Blacks that although the living may forget the dead, the dead themselves never do. Every spring they hold a remembrance service for the dead who have been forgotten by the living.

Brian Lee, 'Empty Pears', first published in *A Fifth Poetry Book*. © Brian Lee 1985. Reprinted by permission of the author.

Alison Leonard, 'Bronwen and the Crows'. © 1992 Alison Leonard. Reprinted by permission of the author.

Douglas Livingstone, 'Pteranodon' reprinted in *Sjambok* (OUP, 1964).

Edward Lowbury, 'Prince Kano' from *Selected and New Poems* (1990). Reprinted by permission of the author and Hippopotamus Press.

Geraldine McCaughrean, 'The Great Swallowing Monster'. © 1992 Geraldine McCaughrean. Reprinted by permission of the author.

Jan Mark, 'No-Good Claus'. © 1992 Jan Mark. Reprinted by permission of Murray Pollinger Ltd.

Patricia Miles, 'Exit' from *The Methuen Book of Strange Tales*, ed. Jean Russell (Methuen Children's Books, 1980). Reprinted by permission of David Higham Associates Ltd.

Susan Price, 'My Great Grandfather's Grave-Digging' from *Here Lies Price* (Faber, 1987). Reprinted by permission of Faber and Faber Ltd. and A. M. Heath & Co. Ltd.

Alison Prince, 'Timmy' from *Haunted Children* (Methuen, 1982). Reprinted by permission of Methuen Children's Books.

Robert Scott, 'Dare You', 'The Slitherydee' and 'Bush Lion'. © 1992 Robert Scott. Reprinted with permission. 'Bush Lion' is retold from a traditional Khoikhoi folktale from the Kalahari region of Southern Africa.

Ian Serraillier, 'The Visitor', first published in *A Second Poetry Book* , ed. John Foster (OUP). © 1980 Ian Serrallier. Reprinted by permission of the author.

Catherine Storr, 'Crossing over' from *Cold Marble and Other Ghost Stories* (Faber, 1985). Reprinted by permission of Faber & Faber Ltd. and the Peters Fraser & Dunlop Group Ltd.

Marilyn Watts, 'The Leopard in the Rafters'. © 1992 Marilyn Watts. Reprinted by permission of the author. With many thanks to Janet, Jeremy and Onesmo. This is based on a story told to the author in Tanzania in 1990.

Although every effort has been made to trace and contact copyright holders prior to printing, this has not been possible in some cases. We apologise for any inadvertent errors or omissions and, if notified, the publisher will be pleased to rectify these at the earliest opportunity.

ILLUSTRATIONS

Mike Belshaw, all chapter opening illustrations; Paul Collicutt, *Leopard in the Rafters*; Sarah Govia, *Eyes of the Cat*; Paul Hunt, *Timmy, No-Good Claus*; Chris Mould, *The Slitherydee, Exit*; Mike Nicholson, *Crossing over*; Chris Price *Tiger in the Snow*; Liz Pyle, *Kiitos! Kiitos!* Martin Remphrey, *Slam and the Ghosts, The Shiner, My Great-Grandfather's Grave-Digging*; Rachael Ross, *A Change of Aunts*; Mark Stacey, *The Great Swallowing Monster*; Adam Stower, *Wiley and the Hairy Man, Bush Lion, Bronwen and the Crows*; Deryk Thomas, *Prince Kano, The Big Black Umbrella*; Anthea Toorchen, *The Friendly Ghost*; Stephen Wilkin, *The Hawk*.

PHOTOGRAPHS

Chris Honeywell small photos on pages 106/7.
Ander McIntyre vignettes on pages 12/13; main photos p 89 and 91.

CALLIGRAPHY

Chapter heading and cover calligraphy by Christopher Moffett.

All other models, photographs and photo montages for *Dear Jane*, *Camilla*, *Dare You* and *Supermarket* are by the designer.

Cover illustration is by Louise Brierley.

this is the child who came to play
on a rainy, windy, nasty day
and said BOO to the ghost who groaned in the hall
and SCAT to the rat by the mouldy wall
and went down the creaking crumbling stair
into the cellar, cold and bare,
and laughed at the spider, huge and fat,
and brushed off the web where it sat and sat
and opened the box
with the rusty locks
and took the key to the castle